Heritage of European Mathematics

Previously published

Andrzej Schinzel, Selecta,
Volume I: Diophantine Problems and Polynomials;
Volume II: Elementary, Analytic and Geometric Number Theory,
Henryk Iwaniec, Wladyslaw Narkiewicz, and Jerzy Urbanowicz (Eds.)

Thomas Harriot's Doctrine of Triangular Numbers: the 'Magisteria Magna', *Janet Beery and Jacqueline Stedall* (Eds.)

Hans Freudenthal, Selecta, *Tony A. Springer and Dirk van Dalen* (Eds.)

Nikolai I. Lobachevsky, Pangeometry, *Athanase Papadopoulos* (Transl. and Ed.)

Jacqueline Stedall

From Cardano's great art to Lagrange's reflections: filling a gap in the history of algebra

European Mathematical Society

Author:

Jacqueline Stedall
The Queen's College
Oxford, OX1 4AW
United Kingdom

E-mail: jackie.stedall@queens.ox.ac.uk

2010 Mathematics Subject Classification: 01-02; 01A40; 01A45; 01A50

Key words: Algebra, equations, renaissance, early modern

ISBN 978-3-03719-092-0

The Swiss National Library lists this publication in The Swiss Book, the Swiss national bibliography, and the detailed bibliographic data are available on the Internet at http://www.helveticat.ch.

Contact address:

European Mathematical Society Publishing House
Seminar for Applied Mathematics
ETH-Zentrum FLI C4
CH-8092 Zürich
Switzerland

Phone: +41 (0)44 632 34 36
Email: info@ems-ph.org
Homepage: www.ems-ph.org

Typeset using the editor's TEX files: I. Zimmermann, Freiburg
Printing and binding: Druckhaus Thomas Müntzer GmbH, Bad Langensalza, Germany
∞ Printed on acid free paper
9 8 7 6 5 4 3 2 1

Contents

Introduction

This book is a quest to understand the transition from the traditional algebra of equation-solving in the sixteenth and seventeenth centuries to the emergence of 'modern' or 'abstract' algebra in the mid nineteenth century. The former was encapsulated in Girolamo Cardano's *Artis magnae, sive, de regulis algebraicis* (*Of the great art, or, on the rules of algebra*), a book commonly known then and now as the *Ars magna*, in 1545. The latter developed out of ideas inspired to a great extent by a seminal paper written by Joseph-Louis Lagrange in the early 1770s, his 'Réflexions sur la résolution algébrique des équations' ('Reflections on the algebraic solution of equations').[1] But what of the two centuries between? When Lagrange embarked on his lengthy 'Réflexions' in the autumn of 1771 he wrote:[2]

> *A l'égard de la résolution des équations litérales, on n'est gueres plus avancé qu'on ne l'étoit du tems du Cardan qui le premier a publié celle des équations du troisieme & du quatrieme degré.*
>
> With regard to the solution of literal equations, we are hardly any more advanced than at the time of Cardano, who was the first to publish that of equations of third and fourth degree.

Most of what follows in this book is essentially an investigation of that claim.

In one sense Lagrange was right: Cardano in 1545 had published rules for solving cubic and quartic equations. Although later writers had added several clarifications and refinements, none had succeeded in working out better or more generally applicable methods. As for fifth or higher degree equations, there was no reason to suppose that they would not in the end yield to similar solution algorithms but, except in a few special cases, there had been no progress in finding them.

In another sense, Lagrange was wrong. There had been many advances in equation-solving since the time of Cardano, some of them small and isolated, others of major significance. In the sixteenth century there had been no general 'theory of equations', only a collection of piecemeal methods and results. By the eighteenth century, however, and in particular during the 1760s, it could finally be said that a theory was beginning to emerge. This was a trend that Lagrange himself, with his keen sense of the history of mathematical thought, both recognized and confirmed in his 'Réflexions'. By examining in depth the writings of his predecessors Lagrange was able not only to generalize old results but to discover new approaches, and to establish the theory on fresh foundations. By the end of his lengthy investigation he was able to write something that to Cardano would surely have seemed inconceivable: that the theory of solving equations reduced to a calculus of combinations, or permutations, of their roots:[3]

[1]Cardano 1545; Lagrange (1770) [1772] and (1771) [1773]. For the double dating system used for articles cited in this book see the note in the bibliography.

[2]Lagrange (1770) [1772], 135.

[3]Lagrange (1771) [1773], 235.

> *Voilà, si je ne me trompe, les vrais principes de la résolution des équations, & l'analyse la plus propre á y conduire; tout se réduit, comme l'on voit, à une espece de calcul des combinations.*
>
> Here, if I am not mistaken, are the true principles of solving equations, and the most correct analysis to lead there; all of which reduces, as one sees, to a kind of calculus of combinations.

The hitherto untold story of the slow and halting journey from Cardano's solution recipes to Lagrange's sophisticated considerations of permutations and functions of the roots of equations is the theme of this present book.

As Lagrange was the first to acknowledge, his ideas rested on work that had been carried out by a number of people during the preceding two centuries. Nevertheless, later writers have continued to perceive the hundred and twenty years before Lagrange as an unfortunate gap in the history of algebra, a period during which little of any importance happened. Luboš Nový, for example, in his *Origins of modern algebra* (1973) recognized Descartes as a major figure but deemed him to have few successors:[4]

> From the propagation of Descartes' algebraic knowledge up to the publication of the important works of Lagrange, Vandermonde and Waring in the years 1770–1, the evolution of algebra was, at first glance, hardly dramatic and one would seek in vain for great and significant works of science and substantial changes.

A few lines later Nový qualified this statement by allowing that over this period algebra gained a new status as the 'language of mathematics', but he nevertheless continued to disregard specific changes or achievements.

Nový can be excused to some extent because the main focus of his text was algebra from a later period, 1770 to 1870. The same cannot be said of B L van der Waerden whose *A history of algebra from al-Khwārizmī to Emmy Noether* (1980) was supposed to offer a complete history of the subject, yet he jumped from Descartes in 1637 straight to Waring, Vandermonde, and Lagrange in the 1770s in the turn of a page, without even a nod towards the lost time in between.[5] Similarly Morris Kline in his 1200-page *Mathematical thought from ancient to modern times* (1972) presented his version of the theory of equations in the seventeenth century in a little under three pages, and in the eighteenth century before Lagrange in just one.[6]

More recently, Isabella Bashmakova and Galina Smirnova in *The beginnings and evolution of algebra* (2000) identified the creation of the theory of equations in the seventeenth and eighteenth centuries as one of the five key stages in the development of algebra, but devoted no more than half a dozen pages to the entire period from Descartes up to Lagrange.[7] Further, Bashmakova and Smirnova, like Kline before

[4] Nový 1973, 23.

[5] Van der Waerden 1980, 75–76.

[6] Kline 1972, I, 270–272; II, 600.

[7] Bashmakova and Smirnova 2000, 94–98, 100–102.

them, present disjointed results from Viète, Descartes, or Euler without any connecting historical or mathematical threads, so that we see only empty spaces between them.

Popular textbooks and general histories have tended to follow much the same pattern.[8] Meanwhile a recent spate of books on the origins of group theory offer similarly brief and somewhat random accounts of progress after Cardano but before Lagrange. Perhaps the most succinct statement comes from Mark Ronan: 'After these successes with equations of degrees 3 and 4, the development stopped.'[9]

This assertion from Ronan, like that from Nový quoted above, betrays a view that mathematics somehow progresses only by means of 'great and significant works' and 'substantial changes'. Fortunately, the truth is far more subtle and far more interesting: mathematics is the result of a cumulative endeavour to which many people have contributed, and not only through their successes but through half-formed thoughts, tentative proposals, partially worked solutions, and even outright failure. No part of mathematics came to birth in the form that it now appears in a modern textbook: mathematical creativity can be slow, sometimes messy, often frustrating.

This book attempts to capture something of the reality of mathematical invention by inviting the reader to follow as closely as possible in the footsteps of the writers themselves. That is to say, the reader is encouraged to put aside modern preconceptions and to approach the problems addressed in this book in the same spirit as the original authors, in the same mathematical language, and with the assumptions, and techniques that were then available. To a modern mathematician, trained to set up careful definitions and rigorous proofs, this may seem somewhat frustrating. The purpose of this book, however, is not to account for modern theory by recourse to historical material, but rather to work from the other direction, to understand how and in what form new ideas began to emerge, by following the historical threads that led to them, without either the benefits or prejudices of hindsight. Inevitably, of course, the ideas and themes we choose to focus on are likely to be those that we know to have been significant later, but the aim is to see them first and foremost from the perspective of their own time.

Internalizing the language, assumptions, and techniques of seventeenth- or eighteenth-century mathematical writers is not easy without immersion in mathematical texts of the period. To help the reader appreciate earlier styles of writing, notation has been left intact as far as possible; where it has been modernized for ease of understanding the original version is given in footnotes. Similarly, where sixteenth-century mathematical Latin has been translated into modern English, the original text is provided for comparison so that readers can see for themselves how much has been lost or gained in translation. On the whole this has not been done for eighteenth-century Latin or French, which in general translate fairly smoothly into English, except where particular words or phrases carry a force of meaning in the original that does not come over well in translation.

---

[8]See, for example, Struik 1954, 114–116, 134; Stillwell 1989, 93–96; Katz 2009, 404–414, 468–473, 671–673.

[9]Ronan 2006, 19; see also Livio 2005, 79–83; Derbyshire 2006, 81–108; Stewart 2007, 75. Du Sautoy 2008 has no references at all for this period.

Assumptions made by past writers can be hard to identify because they were so often just the mathematical common knowledge of the day. Hardly any of the authors featured in this book, for instance, ever specified what numbers could or could not be used as coefficients of equations. At a time when all teaching on equations relied on worked examples, equations were usually given easy integer coefficients, but that does not mean the methods or results were not thought to apply more generally. For Cardano, whose arithmetical world contained integers, fractions, and surds, we can deduce from certain of his statements that he assumed the coefficients of his polynomials to be integers or fractions only, but he never actually said so. After more general notation had been introduced, a distinction was made between 'numerical' and 'literal' coefficients, but still without specifying what kind of numbers the literal coefficients might represent. Such silence persisted into the eighteenth century, by which time literal coefficients could stand not only for numbers but for other algebraic expressions; one usually knows what was intended only from the particular context. It is probably safe to say that where the coefficients stood for numbers, those numbers were, as in Cardano's day, thought to be integers or rationals but in any case they were certainly real: there was no hint of complex coefficients in the eighteenth-century literature on equation-solving.

As for techniques, the modern reader will undoubtedly frequently see shortcuts and better notation that would save many pages of tedious writing. It is a little puzzling, for example, that Lagrange never resorted to some kind of subscript notation instead of running so many times through the alphabet. It is worth recalling, however, that when everything had to be laboriously written or copied by hand there can have been little time for re-writing, correcting, or polishing. In any case, we are not here to mark authors' work with 'could have done better' but to follow what they actually did. I have attempted to point out errors where they invalidate a result that at the time was thought to have been proved, or where they are likely to hinder the reader's understanding, but for the most part the mathematics has been presented in the way it was originally written.

This book is in three parts. Part I offers an overview in three chapters of the period from Cardano (1545) to Newton (1707); here the material is presented chronologically, with explanatory commentary either where the ideas are somewhat obscure in the original (as for Cardano and Viète) or where they are little known (as for Harriot). Part II covers the period from Newton (1707) to Lagrange (early 1770s); by now developments in equation-solving emerged not from relatively isolated texts following one another at irregular intervals, but from a number of different strands of thought which from time to time disappeared or resurfaced, and which often overlapped with each other. For this reason Part II has been arranged by themes, which though roughly chronological in their ordering are not strictly so. Part III is a short account of the dissemination and aftermath of the discoveries made in the 1770s.

Acknowledgements. The research for this book was carried out with the help of funding from the Leverhulme Trust, which has done much to support new initiatives in the history of mathematics in Britain in recent years. For institutional support and many friendships I warmly thank the Provosts, past and present, and the Fellows of The Queen's College, Oxford. I am also grateful to the Mathematical Institute, Oxford. Archival research was carried out at the Berlin-Brandenburgische Akademie der Wissenschaften (BBAW) in Berlin. For both criticism and advice I thank Peter Neumann, who read the first draft of this book with his usual meticulous attention, and also the three referees later appointed by the EMS; I am particularly grateful to the one who returned eight pages of encouraging and perceptive comments and hope that a slippage of anonymity will one day enable me to thank him or her personally. Finally, as I expected from past experience, working with the series editor, Manfred Karbe, and the production editor, Irene Zimmermann, has beeen nothing but a pleasure.

Characters in order of appearance

Joseph-Louis Lagrange (1736–1813)
Girolamo Cardano (1501–1576)
Scipione del Ferro (1465–1526)
Niccolò Tartaglia (1500–1557)
Ludovico Ferrari (1522–1565)
Rafael Bombelli (1526–1572)
Simon Stevin (1548–1620)
François Viète (1540–1603)
Thomas Harriot (1560–1621)
Albert Girard (1595–1632)
René Descartes (1596–1650)
Florimond de Beaune (1601–1652)
Jan Hudde (1628–1704)
François Dulaurens
John Wallis (1616–1703)
John Collins (1625–1683)
James Gregory (1638–1675)
Ehrenfried Walter von Tschirnhaus (1651–1708)
Wilhelm Gottfried Leibniz (1646–1716)
Isaac Newton (1642–1727)
Isaac Barrow (1630–1677)
Walter Warner (1563–1643)
Nicolas Mercator (1620–1687)
John Pell (1611–1685)
Michel Rolle (1652–1719)
Gerard Kinckhuysen (1625–1666)
Colin Maclaurin (1698–1746)
George Campbell (–1766)
Jean Paul de Gua de Malves (1713–1785)
Johann Andreas von Segner (1704–1777)
James Stirling (1692–1770)
Leonhard Euler (1707–1783)
John Colson (1680–1760)
Abraham de Moivre (1667–1754)
Étienne Bezout (1739–1783)
Gabriel Cramer (1704–1752)
Edward Waring (1736–1798)
Alexandre-Théophile Vandermonde (1735–1796)
Paolo Ruffini (1765–1822)
Augustin-Louis Cauchy (1789–1857)
Niels Henrik Abel (1802–1829)
Évariste Galois (1811–1832)

Part I

From Cardano to Newton: 1545 to 1707

Chapter 1
From Cardano to Viète

When Lagrange in his 'Réflexions' in 1771 claimed that there had been scarcely any advance in solving equations since the time of Cardano, he was looking back to results that had first been published in Cardano's *Ars magna* in 1545. Like Lagrange, we too will start from Cardano, but with a different motive. Lagrange saw Cardano's discoveries as an end beyond which no-one had passed; we will regard them instead as a beginning. Historically this was certainly the case: the *Ars magna* set the agenda for the study of equations for the remainder of the sixteenth century and beyond. It will later become apparent that many of the themes and leitmotifs of eighteenth-century equation theory made their first appearance in its pages. A full appraisal of Lagrange's claim therefore requires an understanding first of all of the *Ars magna* itself.

Cardano himself, however, also worked within a pre-existing mathematical context, the world of cossist algebra. This was the algebra derived essentially from al-Khwārizmī's 'Al-jabr w'al-muqābala', written in Baghdad around 825 AD, which gave rules for solving various kinds of quadratic equation. The different cases arose from the convention that all terms were expressed positively and the fact that any one of the three terms 'square', 'thing', or 'number' might or might not appear. Thus, there were six possibilities, starting with 'squares equal to things' (in modern notation $ax^2 = bx$) and ending with 'things plus numbers equal to squares' ($bx + c = ax^2$). Al-Khwārizmī's treatise was rendered into Latin in the twelfth century, but Islamic algebra became more widely known through the fifteenth chapter of the 'Liber abaci' of Leonardo Pisano (Fibonacci) of 1202, and later Italian abacus texts.[1]

The term 'cossist' derives from the word *cosa* used by Italian writers for the unknown 'thing'. The common manuscript abbreviations *co*, *ce*, *cu*, for *cosa*, *census* (square), and *cubus* (cube), were eventually replaced by single letters: R for things (*res*) or roots (*radices*), Z for squares, C for cubes, and sometimes N for numbers. In early printed algebra texts, published in Italy, Germany, France, Spain, and England, each author devised his own version of the notation, but the rules taught were essentially the same as those given by al-Khwārizmī 800 years earlier.

Attempts at cubic equations in such texts were infrequent and were often based on futile applications of the rules for quadratics. Paolo Gerardi in 1328, for example, claimed he had solved the equation 8 *cubi sono iguali a* 9 *ciensi e a* 4 *cose e a* 12 *in numero* (in modern notation $8x^3 = 9x^2 + 4x + 12$) by adding 4 *cose* to 12 to make 16 and then treating the equation as a quadratic. He was not alone in propagating such methods.[2] Eventually, Luca Pacioli in his *Summa de arithmetica* of 1494, the first printed mathematical treatise to include algebra, understood that this would not

[1] Leonardo 2002, 531–615; see also van Egmond 1978, 1983; Franci and Rigatelli 1985; Høyrup 2007; Céu Silva 2008.

[2] Van Egmond 1978; Céu Silva, 2008.

do, and declared that only equations that were reducible to the six standard cases were solvable.[3] Thus, he argued, 'square of squares' plus 'squares' equal to 'numbers' could be handled, but 'squares of squares' plus 'squares' equal to 'things' could not.[4] Within just a few years of Pacioli's death, however, correct solutions to certain types of cubic equation were discovered and began to be passed around by word of mouth between a tiny handful of north Italian practitioners. It was here that Cardano entered the story.

Cardano and the *Ars magna*, 1545

Girolamo Cardano was born in 1501 in Pavia in northern Italy.[5] He studied at the universities of Pavia and Padua, and trained in medicine, which he practised for most of his life. In the early 1530s he moved to Milan where he also began to teach mathematics; his first book on the subject, his *Practica arithmetice*, was published in 1539. Cardano took up the chair of medicine at Pavia from 1543 and held it until 1560, at which point his life was severely disrupted by family troubles. Two years later he returned to medicine, now at Bologna. His academic career came to an end, however, in 1570 when he was imprisoned for heresy (for casting the horoscope of Christ), and he was afterwards forbidden to teach. He spent the remaining years of his life in Rome where he died in 1576.

A man of broad learning, Cardano wrote a great many books, on medicine, philosophy, science, and mathematics.[6] One of them was the *Liber de ludo aleae*, one of the earliest mathematical treatises on games of chance, written in or after 1564 but not published until the seventeenth century.[7] The *Ars magna* was the tenth of a set of fourteen books on mathematical subjects, of which the first nine dealt with various aspects of arithmetic, and the last four with geometry. Most of these are known to have been written but not all of them have survived.[8] We also have Cardano's *Opus novum de proportionibus numerorum* of 1570, which includes what was advertised as a revised and augmented edition of the *Ars magna*, though the changes from the first edition are slight.

Cardano was already well versed in the algebra of the early sixteenth century when he wrote his *Practica arithmetice* in the 1530s. In his first chapter he defined 'named numbers' (*numerus denominatus*): roots, squares, cubes, and so on, and claimed that although these were numbers 'only in resemblance' (*solum per similitudinem*),[9] the

[3] *Altramente che in questi .6. discorsi modi non e possibile alcuna loro equatione.* [Other than in these 6 ways discussed, it is not possible [to solve] any equation.] Pacioli 1494, 144v.

[4] Pacioli 1494, 149.

[5] For Cardano's biography see Cardano 1931; Ore 1953.

[6] Cardano's *Opera omnia* published in Leiden in 1663 consists of 10 volumes. Volumes I to III contain mainly philosophical writings and the *Liber de ludo aleae*; Volume IV contains 'Arithmetica, geometrica, musica' and Volume V contains 'Astronomia, astrologica, onirocritica'; Volumes VI to IX contain writings on medicine; Volume X consists of 'miscellanea', including more mathematics.

[7] Cardano 1663, I, 262–276; Ore 1953; Bellhouse 2005.

[8] For a list of Cardano's mathematical treatises see Cardano 1663, I, 66 and 74.

[9] *Numerus denominatus est, ille qui solum est numerus per similitudinem, veluti Radix, census, cubus, & tales.* [A named number is that which is a number only in resemblance, like a root, square, cube, and suchlike.] Cardano 1663, IV, 14.

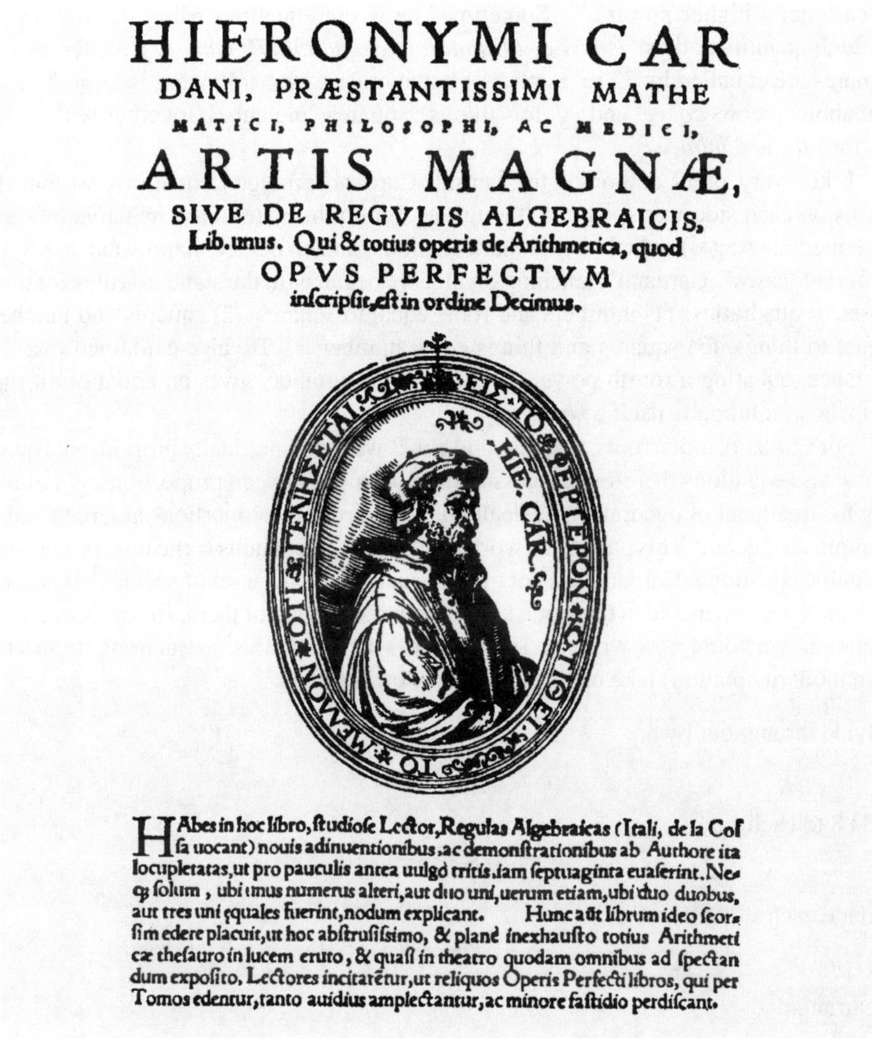

HIERONYMI CAR
DANI, PRÆSTANTISSIMI MATHE
MATICI, PHILOSOPHI, AC MEDICI,
ARTIS MAGNÆ,
SIVE DE REGVLIS ALGEBRAICIS,
Lib. unus. Qui & totius operis de Arithmetica, quod
OPVS PERFECTVM
inſcripſit, eſt in ordine Decimus.

HAbes in hoc libro, ſtudioſe Lector, Regulas Algebraicas (Itali, de la Coſ ſa uocant) nouis adinuentionibus, ac demonſtrationibus ab Authore ita locupletatas, ut pro pauculis antea uulgò tritis, iam ſeptuaginta euaſerint. Neq; ſolum, ubi unus numerus alteri, aut duo uni, uerum etiam, ubi duo duobus, aut tres uni æquales fuerint, nodum explicant. Hunc aũt librum ideo ſeorſim edere placuit, ut hoc abſtruſiſsimo, & planè inexhauſto totius Arithmeticæ theſauro in lucem eruto, & quaſi in theatro quodam omnibus ad ſpectandum expoſito, Lectores incitarẽtur, ut reliquos Operis Perfecti libros, qui per Tomos edentur, tanto auidius amplectantur, ac minore faſtidio perdiſcant.

A new understanding of equations (1): Cardano's *Ars magna* (1545), containing treatments of cubics, quartics, and transformations.

usual operations of arithmetic could be carried out on them just as for as the three other kinds of number in his universe: integers, rationals, and surds. Cardano's chapters 48 to 51 were specifically devoted to algebra, which for him was concerned with relationships between numbers and unknown 'things', their squares and cubes, and occasionally higher powers.[10] Sometimes he wrote equations using the full names of such quantities, thus: *cubus & 11 quadrata aequantur 72 numero* ('a cube and 11 squares are equalled by 72 in numbers') but at other times, following Pacioli, he used the abbreviations *co*, *ce*, and *cu*, for 'things', squares, and cubes, together with $\tilde{p}$. and $\tilde{m}$. for *plus* and *minus*.

Like every other author of the period, Cardano arranged equations so that the terms on each side were positive (though he did not hesitate to use negative terms in intermediate stages), which meant that equations of each degree manifested as several different 'cases'. Cardano's teaching on algebra began with the standard rules for three cases of quadratics: (1) numbers and roots equal to squares, (2) squares and numbers equal to things, (3) squares and things equal numbers.[11] He also explained that, for instance, equating a fourth power to squares plus numbers gives an equation of type (1) whose solution is itself a square.

For Cardano, units, roots, squares, and cubes were geometrically proportional quantities, and equations therefore expressed relationships between proportions.[12] Following his treatment of quadratics, he dealt with properties of proportions at length and in complicated detail. This part of his work also, however, contains some interesting work on cubic equations. Cardano did not give a general rule, but a set of special cases, each of which was amenable to the same kind of treatment. One of them, for instance, is the equation we would now write as $3x^3 = 15x + 6$. Cardano's instructions (translated into modern notation) take us through the following steps.[13]

Divide throughout by 3:

$$x^3 = 5x + 2.$$

Add 8 to each side:

$$x^3 + 8 = 5x + 10.$$

Divide each side by $x + 2$:

$$x^2 - 2x + 4 = 5.$$

Rearrange:

$$x^2 = 2x + 1.$$

---

[10] *In algebra considerantur denominationes videlicet numerus, res vel radix; census & cubus, & census census, & reliquo dicta in primo capitulo.* [In algebra we examine what are called number, thing or root, square and cube, and square-square, and the rest, as given in the first chapter.] Cardano 1663, IV, 71.

[11] *numerus & radix aequantur censibus* [...] *census & numerus aequantur rebus* [...] *census & res aequantur numero*. Cardano 1663, IV, 72.

[12] *His visis scire quod numerus co. ce. cu. sunt semper apud algebra continuae proportionalia.* [From this it is understood that number, things, squares, cubes in algebra are always proportional quantities.] Cardano 1663, IV, 77.

[13] *cubis 3. aequales 15. co. $\tilde{p}$ 6.* Cardano 1663, IV, 81.

By the usual rules for quadratics, solve for the positive root:

$$x = 1 + \sqrt{2}.$$

This example, and others in the same section, rely on division of $x^3 \pm a^3$ by $x \pm a$ and so show an understanding of polynomial division. Even after he later discovered general rules for cubics and quartics, Cardano was always keen to display special cases that could be handled by 'short cuts', and there are numerous examples of 'special' cubics and quartics in the *Ars magna*.

The earliest known discoverer of a general rule for cubics, of the particular form $x^3 + cx = d$, was Scipione del Ferro, in Bologna around 1520. In the late 1530s Niccolò Tartaglia of Brescia independently rediscovered the method when he was challenged by one of del Ferro's pupils, Antonio del Fior, to answer a set of questions that all led to cubic equations of the above type. In 1539 Cardano persuaded Tartaglia to teach him the method and Tartaglia gave it to him in the form of a verse (which rhymes much more nicely in Italian):[14]

When the cube with the things next after
Together equal some number apart
Find two others that by this differ
And this you will then keep as a rule
That their product will always be equal
To the cube of a third of the number of things
The difference then in general between
The sides of the cubes subtracted well
Will be your principal thing.

For an equation of the form $x^3 + cx = d$, the verse instructs us to find two numbers which we may call u and v, such that $u - v = d$ and $uv = (\frac{c}{3})^3$. The required solution will then be $x = \sqrt[3]{u} - \sqrt[3]{v}$. It is easily checked that this expression for x does indeed satisfy the equation.

According to his own account, Cardano felt free to publish this when he found that Tartaglia was not the first to have discovered it, and it became one of the central teachings of the *Ars magna*. Tartaglia's reaction was understandably bitter, even though Cardano more than once gave him credit. For Cardano, however, Tartaglia's rule was just one of several new ideas in the book, or rather, a starting point that led him into a multitude of new discoveries.

The *Ars magna* is a treasury of rules, methods, observations, insights, and special cases, but the reader has to work hard for them. The ordering of the material is often haphazard, with many diversions and repetitions, and Cardano's language is verbose, dense, and sometimes ambiguous. One of the greatest difficulties for the modern

[14]*Quando chel cubo con le cose apresso / Se aguaglia à qualche numero discreto / Trovan dui altri differenti in esso / Dapoi terrai questo per consueto / Ch'el lor' produtto sempre fia equale / Al terzo cubo delle cose neto / El residuo poi suo generale / Delli lor lati cubi ben sottratti / Varra la tua cosa principale.* Tartaglia 1546, 124.

reader is the absence of symbolic notation. Every equation or rule becomes a sentence, sometimes of considerable length, which the reader must hold in mind from beginning to end. The rearrangement of an equation or the application of a rule leads to another sentence, which must be referred back step by step to the previous one. Indeed much of the book is taken up with complex and unmemorable verbal instructions, which become trivially easy once they are re-written in modern notation. Most difficult of all are the passages where Cardano worked with two unknown quantities, both denoted by the same word *positio* or 'supposed quantity'. For most readers the distinctions and relationships between the two unknowns are almost impossible to hold in mind without reverting to some kind of symbolism, and one cannot but admire Cardano's mental agility in working without it. For ease of reading and comprehension, we will take an easier route and for the remainder of this section we will clothe Cardano's findings in modern notation.

There are 40 chapters in the *Ars magna*. Those particularly concerned with solving equations are the following.

Chapter 1	Comments on equations with more than one root
Chapters 1, 3, 4, 6, 37	Comments on negative, surd, and complex roots
Chapters 2–4, 24	General rules for simplifying equations
Chapter 5	Solution of quadratic equations
Chapter 6	Some new methods
Chapters 7–8, 25–26, 40	Some special cases
Chapters 11–23	Solution of cubic equations
Chapter 29	Simultaneous linear equations
Chapter 30	Finding an approximate root
Chapters 31–36, 38	Problems leading to polynomial equations
Chapter 39	Solution of quartic equations

Chapter 1 consists largely of rules for the number of positive or negative roots of cubics, and is a summary of material treated at greater length in Chapters 11–23. It reads like a later addition to the book and so we will put it aside for now and return to it in the context of Cardano's comments elsewhere in the *Ars magna* on the number and nature of roots of equations. We therefore begin here with Chapters 2 to 5.

Chapters 2 to 5, almost certainly written before the present Chapter 1, contain the kind of rules and instructions that by 1545 were commonplace in elementary algebra texts. Cardano pointed out, for example, that equations of the form $d = x^4 + cx^2$ and $d = x^6 + cx^3$ are both related to the simple quadratic $d = x^2 + cx$, with a long list of similar examples. He instructed that one should simplify an equation by dividing through by common powers of x and by the leading coefficient, thus reducing the polynomial to one that we would now describe as 'monic' (with leading coefficient 1). He then gave the usual rules for solving the three standard types of quadratic equation: (1) $d = x^2 + cx$, which he called *nuquer* ('nqr': numbers equal to square plus roots); (2) $x^2 = cx + d$, or *querna* ('qrn': square equal to roots plus numbers);

(3) $cx = x^2 + d$, or *requan* ('rqn': roots equal to square plus numbers). Each rule was accompanied by geometric demonstrations of the 'cut-and-paste' variety, which offer visual justifications for the verbal rules.

It was in Chapter 6 that Cardano began to break new ground, and he did so with a very simple problem.[15] Suppose we want to find two numbers whose sum is the square of one of them and whose product is 8. In other words, writing x and y for the two unknown numbers, we have the equations $x + y = x^2$ and $xy = 8$. The substitution $y = 8/x$ gives the equation

$$x^2 + 8 = x^3.$$

If we make a different substitution, however, namely $x = 8/y$, we arrive at

$$8y + y^3 = 64.$$

For Cardano these were different kinds of equation: the first belongs to the type 'cube equal to squares and numbers' whereas the second is of the type 'cube and roots equal to numbers'. Clearly, however, their solutions are related by the simple transformation $x = 8/y$ or $y = 8/x$. Since Cardano knew how to solve the second kind of equation (by Tartaglia's rule) he could also see how to solve the first. This is the first example we have in the history of algebra of the transformation of an equation by an operation on the roots. It is of fundamental importance. Cardano's optimism at this point shines through in his writing: 'Transform problems that are by some ingenuity understood to those that are not understood', he wrote, 'and there will be no end to the discovery of rules'.[16]

Now Cardano had an insight into cubics of the types 'cube, thing, number' and 'cube, square, number' and could give specific solution recipes for all of them (Chapters 11–16). The first rule, for 'cubes and things equal to a number', arose directly from Tartaglia's method, but became commonly known as 'Cardano's rule':[17]

> Raise a third part of the number of things to a cube, to which you add the square of half the number of the equation, and take the root of all of it, that is the square root, which you put twice, and to one you add half of the number, which you multiplied by itself, and from the other you subtract

[15] *duos inuenias numeros, quorum aggregatum aequale fit alterius q̃drato, & ex uno in laterum ducto, producatur 8, una enim uia peruenies ad 1 cubum aequalem 1 q̃drato p: 8, alia, ad 1 cubum p: 8 rebus, aequalem 64, hac igitur inuenta aestimatione, si diuiseris 8 per eam, prodibit reliqua equatio, ex qua in capituli illius cogitationem perueni.* [Find two numbers whose sum is equal to the square of one, and such that one multiplied by the other produces 8, for one way you come to 1 cube equal to 1 square plus 8, and the other to 1 cube plus 8 things equal to 64, thus having found the root if you divide 8 by it, it will produce the other solution, from which I came to knowledge of the rule for that one.] Cardano 1545, 15v–16; 1663, IV, 235; 1968, 51–52.

[16] *Quaestiones igitur alio ingenio cognitas ad ignotas transfer positiones, nec capitulorũ inuentio finem est habitura.* Cardano 1545, 16; 1663, IV, 235; 1968, 52.

[17] *Deducito tertiam partem numeri rerum ad cubum, cui addes quadratum dimidij numeri aequationis, & totius accipe radicem, scilicet quadratam, quam [g]eminabis, uniq̃ dimidium numeri quod iam in se duxeras, adijcies, ab altera dimidium idem minues, habebisq̃ Binomium cum sua Apotome, inde detracta R cubica Apotome ex R cubica sui Binomij, residuũ quod ex hoc relinquitur, est rei estimatio.* Cardano 1545, 30; 1663, IV, 250; 1968, 98–99.

> the same half, and you will have a binome with its apotome, whence when the cube root of the apotome has been subtracted from the cube root of its binome, the difference that remains, that is the solution.

In other words, for an equation of the form $x^3 + cx = d$, the required solution is

$$x = \sqrt[3]{\sqrt{\frac{c^3}{27} + \frac{d^2}{4}} + \frac{d}{2}} - \sqrt[3]{\sqrt{\frac{c^3}{27} + \frac{d^2}{4}} - \frac{d}{2}}.$$

If this rule is applied to the case $x^3 = cx + d$, however, the square roots will contain the term $\frac{d^2}{4} - \frac{c^3}{27}$, which becomes negative if $\frac{c^3}{27} > \frac{d^2}{4}$. For Cardano, roots of negative quantities made no sense. In August 1539 he had written to Tartaglia seeking clarification but had not received it.[18] In the *Ars magna* he skirted around the problem: in Chapter 12, on 'The cube equal to the first power and number' he instructed his readers that whenever the cube of one-third of the coefficient of the linear term was greater than the square of one-half of the numerical term (in modern notation $\frac{c^3}{27} > \frac{d^2}{4}$) they should try a geometric approach or else turn to Chapter 25, where they would find particular rules for avoiding the difficulty (one of them is the technique outlined above for the equation $x^3 = 5x + 2$).

The equations that lead to this impasse (in modern terms $x^3 = cx \pm d$ with $\frac{c^3}{27} > \frac{d^2}{4}$) have three real roots, but Cardano's rule appears to yield a complex or 'impossible' root. As Bombelli found later, this root is in fact a sum of complex conjugates, which means that the imaginary parts cancel out. To find the conjugates, however, one has to take cube roots of complex numbers, and except where these can be seen by inspection one is led straight back to the original cubic equation. In short, Cardano's rule did not seem to be helpful. Cardano himself devoted a great deal of energy to exploring this case further in a treatise with the untranslatable title 'De regula aliza' in his *Opus novum* of 1570, but clearly he, like later writers, thought that he had failed find a general rule that was valid for all cubics.[19]

After the cases in which either 'squares' or 'things' were missing, there remained only the apparently more general forms of cubic equations involving all four quantities, 'cubes, squares, things, and numbers', but Cardano saw how to handle these too. Consider the equation

$$x^3 + 6x^2 + 20x = 100. \tag{1}$$

[18]Tartaglia 1546, IX, 125v–127v.

[19]Cardano 1663, IV, 377–434. In a letter to Huygens, probably written in September 1675, Leibniz claimed to have proved that Cardano's rule gave a general solution for all cubic equations (*Je croy d'avoir demonstré que les formules de Cardan sont absolument bonnes et generales*). Leibniz's concern was to show that a root composed only of integers and square roots could be elicited by Cardano's rule, even though the latter appeared always to produce cube roots. Take for instance the equation $x^3 - 12x = 9$. Bombelli, following Cardano's method, had reduced the equation to $x^2 + 3 = 3x$ with the positive root $1\frac{1}{2} + \sqrt{5\frac{1}{4}}$, and had assumed that such a root could not be found by Cardano's rule. Leibniz insisted that it could, but did not show how. See Leibniz 1976 (3), I, 277–278.

Now make the substitution $x = y - 2$, where 2 is chosen because it is one-third of 6, the coefficient of x^2. The equation now reduces to

$$y^3 + 8y = 124, \qquad (2)$$

which was one that Cardano could solve. Cardano offered precisely this example and demonstrated it geometrically with a diagram of a cube suitably partitioned.[20] Immediately afterwards he explained that it is always possible to move from equations of type (1) to equations of type (2) and gave separate rules for writing down each coefficient of (2). The crucial quantity is 'one third of the number of squares', which is used to eliminate the square term. Cardano denoted it by 'Tpq̃d' (T[ertia] p[ars] q[ua]d[ratorum]),[21] and instructed that it must be added to or subtracted from the solution to the transformed equation to give the required solution to the original equation.[22]

For Cardano this transformation, which so conveniently removed the square term from any cubic, opened up the solution of cubics in general (Chapters 17–23). A similar transformation can be used for removing the cube term from a quartic, or indeed the second highest term from any polynomial, and although Cardano himself did not use it on higher degree equations, later sixteenth-century writers certainly did. Indeed, the substitutions $y = k/x$ and $y = x \pm k$ rapidly became standard tools of equation solving.

Both before and after his long central section on cubics, Cardano devoted special attention (in Chapters 8, 25) to equations with just three terms, of the form

$$x^n + q = px^m$$

or

$$x^n + px^m = q$$

with $n > m$. Quadratic equations, and cubics of the type 'cube, thing, number' or 'cube, square, number', are special cases of three-term equations, and this may have been what led Cardano to study them. Borrowing the language of proportion, Cardano called the terms 'x^n' and 'q' the extremes and 'x^m' the mean. For equations of the form

$$x^n + q = px^m$$

he gave the following rule:[23]

---

[20] *cubus AB + 6 quadrata, & 20 positiones aequalia 100.* Cardano 1545, 36; 1663, IV, 256; 1968, 121–122.

[21] 3^{m} *partem numeri quadratorum (quam hoc signo, Tpq̃d: demonstramus)* ... [A third part of the number of squares (which we will denote by this sign: Tpq̃d) ....] Cardano 1545, 36; 1663, IV, 256 (but the latter has 'Tpquad'); 1968, 122.

[22] *ut aestimatione inventae addatur aut minuatur Tpq̃d.* [For finding the solution, Tpq̃d is added or subtracted.] Cardano 1545, 36v; 1663, IV, 257; 1968, 124.

[23] *seceris duas partes, ex quarum una in radicum alterius, sumptam secundam naturam denominationis, prouenientis ex diuisione extremae per mediam, & deductam ad naturam ipsius mediae denominationis, fiat numeris aequationis, huc radix ipsa anteq̃ deducetur ad naturam denominationis mediae, est rei aestimatio.* Cardano 1545, 21; 1663, IV, 240; 1968, 68.

> You will cut [the coefficient of the mean] into two parts, of which one times the root of the other, taken according to the nature of the power arising from division of the extreme by the mean, and raised according to the nature of the power of that mean, makes the number of the equation, this root which before being raised according to the nature of the power of the mean, is the solution.

In modern notation we may write this rule as follows: find two numbers, a and b, such that

$$a + b = p$$

and

$$ba^{\frac{m}{n-m}} = q.$$

Then $x = a^{\frac{1}{n-m}}$ will be a solution to the equation. It is easily checked that this is correct.[24] Cardano gave no justification, however, only some well chosen examples in which a and b can be found by inspection. For the equation $10x^3 = x^5 + 48$, for example,[25] the coefficient 10 may be partitioned into 6 and 4, since 6 times $4^{\frac{3}{5-3}}$ is 48. The required root of the equation is then $4^{\frac{1}{5-3}} = 2$. As we will see later, the rationale behind this method became clearer in the work of Viète.

Cardano would have known, of course, that a given polynomial equation might not yield, or at least not easily, to any of the rules he had been able to give so far. In Chapter 30, therefore, he made a brief foray into a numerical method, to cover the cases that 'come about in practice'.[26] His first step is easy enough to follow: he suggests a simple linear interpolation between a value that is too small and another that is too large. The refinements he proposes after that, however, are not based on any comprehensible reasoning, and Cardano does not offer enough examples to make his method clear.

In the penultimate chapter of the book Cardano offered what he perhaps regarded as its crowning glory, a method worked out by his pupil Ludovico Ferrari for solving quartic equations. Cardano did not state a general rule, but gave seven worked examples which make the method clear. To illustrate it, here is his solution of the equation we can write as $x^4 = x + 2$. Here the letter x stands for Cardano's first *positio*, that is, his supposed or unknown quantity. Almost immediately, he introduced a second unknown quantity, also called *positio*, which we will denote by y. In Cardano's exposition, however, the same word is used for both and one can distinguish between them only from context.[27]

---

[24] The solution given by Witmer in Cardano 1968, 68 n 3, is incorrect.

[25] *10 cubi aequantur* $p^o R^o$ + 48. Cardano 1545, 21–21v; 1663, IV, 241. The naming of powers higher than cubes required some ingenuity. Fourth, sixth, eighth, ninth, and higher composite powers could be described as 'square-squares', 'square-cubes', 'square-square-squares', 'cube-cubes', and so on. Prime powers, however, had to be given individual names. The most common way of describing a fifth power was as 'primo relato', hence $p^o R^o$; a seventh power was 'secundo relato', and so on. Such a scheme was set out in Pacioli 1494, 143, and was followed by many later writers.

[26] *Haec regula rerum, quae in usum veniunt, maximã partem amplectitur.* [This rule will embrace most things that come about in practice.] Cardano 1545, 53; 1663, IV, 273; 1968, 182.

[27] *quia igitur additis 2 positionibus p: 1 quadrato numeri quadratorum, ad 1 positionem p: 2, fit totum 2*

To the left-hand side of the equation Cardano added the quantity that in our notation would be written as $2yx^2 + y^2$, thus ensuring that this side of the equation remains a perfect square. Balancing the two sides we therefore have

$$x^4 + 2yx^2 + y^2 = 2yx^2 + x + (2 + y^2). \tag{3}$$

The right-hand side is quadratic in x, and by judicious choice of y can also be made into a perfect square. The condition for this is [28]

$$\frac{1}{4} = 2y^3 + 4y. \tag{4}$$

But this is just a cubic equation in y, of a form that Cardano could solve.

Making use of any value of y that satisfies (4), Cardano could now reduce (3) to an equation between two squares. The left-hand side is the square of

$$x^2 + y, \tag{5}$$

and the right hand side is the square of

$$x\sqrt{2y} + \sqrt{2 + y^2}. \tag{6}$$

Equating (5) and (6), Cardano therefore had

$$x^2 = x\sqrt{2y} + \sqrt{2 + y^2} - y,$$

a straightforward quadratic equation that can be solved in the usual way. The principle of the method is clear: one must first solve a cubic and then use one of its roots to reduce the quartic to a product of quadratics. In practice, this leads to lengthy strings of nested cube and square roots, and one can only wonder at the patience and persistence of Ferrari and Cardano in pursuing the method to its end.

---

positiones numeri quadratorum p; 1 pos. p: 2, p: 1 quadrato numeri quadratorum, et hoc habet radicem, oportet ut quadratum dimidij mediae quantitatis, quae est 1 positio, aequetur ductui extremorum, igitur $\frac{1}{4}$ quadrati, aequabitur quadrato, 2 cuborum p: 4 positionibus numeri prioris, quare abiectis quadratis utrinque, fiet $\frac{1}{4}$ aequalis 2 cubis p: 4 positionibus, et $\frac{1}{8}$ aequalis 1 cubo p: 2 positionibus, quare rei aestimatio est $\mathcal{R}$v: cubica $\mathcal{R}$ $\frac{2075}{6912}$ p: $\frac{1}{16}$ m: $\mathcal{R}$v: cubica $\mathcal{R}$ $\frac{2075}{6912}$ m: $\frac{1}{16}$, hic igitur est numerus quadratorum addendus utrique parti, et duplicatur, et quadratum huius erit numerus addendus ad utramque partem. [Since therefore by the addition of two unknown numbers [of squares] plus a square of the number of squares, to 1 unknown plus two, it will make in all two unknown numbers of squares plus one unknown plus 2, plus a square of the number of squares, and for this to have a root, it must be that the square of half of the mean quantity, which is 1 unknown, is equal to the product of the extremes, therefore $\frac{1}{4}$ of a square will equal 2 cubes plus 4 unknowns of the square of the first number, from which having eliminated squares on both sides, there will be $\frac{1}{4}$ equal to 2 cubes plus 4 unknowns, and $\frac{1}{8}$ equal to 1 cube plus 2 unknowns, whence the solution $\sqrt[3]{\sqrt{\frac{2075}{6912}} + \frac{1}{16}} - \sqrt[3]{\sqrt{\frac{2075}{6912}} - \frac{1}{16}}$, this therefore is the number of squares to be added to each part, and doubled, and the square of it will be the number to be added to each part.] Cardano 1545, 75–75v; 1663, IV, 296; 1968, 243–244.

[28]The quadratic expression $ax^2 + bx + c$ is a perfect square if and only if $\frac{b}{2}^2 = ac$.

Nevertheless, they did so, finding, for example, that a (positive) root of the above equation, $x^4 = x + 2$, is

$$\sqrt{\sqrt[3]{\tfrac{1051}{3456} + \sqrt{\tfrac{2075}{442368}}} + \sqrt[3]{\tfrac{1051}{3456} - \sqrt{\tfrac{2075}{442368}}} + \tfrac{2}{3} - \sqrt[3]{\sqrt{\tfrac{2075}{442368}} + \tfrac{1}{128}} - \sqrt[3]{\sqrt{\tfrac{2075}{442368}} - \tfrac{1}{128}}}$$
$$+\sqrt{\sqrt[3]{\sqrt{\tfrac{2075}{442368}} + \tfrac{1}{128}} - \sqrt[3]{\sqrt{\tfrac{2075}{442368}} - \tfrac{1}{128}}}.$$

For this particular equation Cardano also had another method. Rearranging it as $x^4 - 1 = x + 1$, he could divide both sides by $x + 1$ and so reduce it to $x^3 + x = x^2 + 2$. By the standard method for cubics, this yields a root equal to

$$\sqrt[3]{\sqrt{\tfrac{2241}{2916}} + \tfrac{47}{54}} - \sqrt[3]{\sqrt{\tfrac{2241}{2916}} - \tfrac{47}{54}} + \tfrac{1}{3}.$$

Cardano assumed that these different methods had elicited the same root, and therefore that it must be possible to reduce the first form to the second by manipulation of surds. He had shown earlier in the book how to do this kind of thing, though never with an example as difficult as this, and he did not work it through here. He asserted, however, that completing such a reduction was one of the greatest things the perfection of the human intellect, or rather the human imagination, could achieve.[29]

Finally, we return to Cardano's opening chapter, taken together with some further remarks scattered through the rest of the book on the nature of the roots of equations. Chapter 1 has the title 'On double solutions in certain types of cases'. Cardano began by pointing out that a square number has two roots, one positive and one negative. Thus since the equation $x^4 + 12 = 7x^2$ is satisfied by $x^2 = 4$ or $x^2 = 3$, it has four possible roots: $2, -2, \sqrt{3}, -\sqrt{3}$. At this point he began to call positive roots 'true' and negative roots 'feigned' or 'fictitious'.[30] Although negative roots do not occur later in the book, Cardano was perfectly well able to handle them. Thus, for example, he showed in Chapter 1 that -4 is a root of $x^3 + 16 = 12x$ by correctly evaluating the substitution on both sides of the equation.[31] Imaginary roots, however, were barely on his horizon, and Cardano claimed that the equation $x^3 + 6x = 20$ has no solution other than 2, neither true nor fictitious.[32]

[29] *est ideo complementum in hic operationibus, est quasi extremum, ad quod peruenit perfectio humani intellectus, uel potis imaginationis.* [Therefore the completion of this operation is as though the greatest thing the human intellect, or rather imagination, can arrive at.] (In other words, it is the reduction of the surd forms, rather than the solution of the quartic itself, that Cardano seemed to think was the greatest challenge.) Cardano 1545, 75v; 1663, IV, 297; 1968, 246.

[30] *ficta, sic eñ vocamus, quae debiti est seu minoris.* [Fictitious, thus we call them, where they are owed or less.] Cardano 1545, 3v; 1663, IV, 223; 1968, 11.

[31] *si cubus p̃: 16, aequatur 12 positionibus, estimatio rei est m̃:4, nam 12 res sunt m̃:48, & cubus m̃:4 est m̃:64, cui additio 16 fit m̃:48.* [If a cube plus 16 makes 12 things, the estimated thing is −4, for 12 things are −48, and the cube of −4 is −64, to which the addition of 16 makes −48.] Cardano 1545, 4v; 1663, IV, 223; 1968, 12.

[32] *1 cubus p: 6 positionibus, aequatur 20, rei aestimatio nulla est praeter 2, neque vera neque ficta.* [For a cube plus six unknowns equal to 20, there is no solution besides 2, neither true nor fictitious.] Cardano 1545, 4; 1663, IV, 223; 1968, 11.

On the number of positive or negative roots of cubics in general he was able to go much further. He instructed that for the equation $x^3 + d = cx$ one should calculate $\frac{2c}{3}\sqrt{\frac{c}{3}}$ and compare it with d. If they are the same, he claimed, the equation will have one positive root and one negative; if the first is greater, the equation will have two positive roots and one negative (the latter being equal in absolute value to the sum of the two positive roots); if it is smaller, the equation will have no positive roots but one negative root.[33] In fact he gave rules in this first chapter (and again later) for the number of positive and negative roots for all types of cubic, though in the most general cases ('cubes, squares, things, numbers') the rules, sub-rules, and special conditions multiply alarmingly.

Further, Cardano noted that a fictitious root of $x^3 + d = cx$ is a true root of $x^3 = cx + d$. The modern perception of this is that these two equations transform to one another if x is replaced $-x$. Cardano did not say so explicitly but seems to have had some such transformation in mind, and to have used it repeatedly. Indeed at the end of his chapter he demonstrated geometrically that a true root of

$$x^3 + 10 = 6x^2 + 8x$$

must be a fictitious root of

$$x^3 + 6x^2 = 8x + 10,$$

where again the second equation is obtained from the first by replacing x by $-x$.

As to the total number of roots (which for Cardano meant what we would call *real* roots), here too he was able to make some important general observations. One was that a cubic equation may have either three or one roots, while a quartic may have four, two, or none.[34] Further, he observed that where a cubic equation has three (real) roots, their sum is always the coefficient of the square term.[35]

The analogies between 'cubes, squares, lines, and numbers' in geometry and 'cubes, squares, things, and numbers' in algebra allowed Cardano (like his predecessors) to offer several convincing geometric demonstrations of algebraic procedures. At the same time, such analogies were restrictive: Euclidean geometry deals only with positive magnitudes in three dimensions. This led Cardano to assert that it would be foolish to go beyond cubes because in nature such a thing is impossible,[36] (though he did deal with quartics, as 'squares of squares'). Further, negative roots (or sides) of squares

[33] *vide an ex duabus tertijs numeri Rerum in radicem tertiae partis eiusdem numeri fiat ducendo, numero propositus aut maior, aut minor.* [Look at the number that arises from multiplying two-thirds of the number of things by the square root of a third of the same, whether it is the proposed number or larger or smaller.] Cardano 1545, 4; 1663, IV, 223; 1968, 11–12.

[34] Cardano's note on this subject begins: *Notum est autem ex hoc, quod capitula quaedam habent duas, quaedam unam aestimationem, ...* [It is to be noted, moreover, that certain cases have two solutions, certain others one ....] In the light of the rest of the paragraph, where cases are listed explicitly, 'duas' here is almost certainly a misprint for 'tres'. Cardano 1545, 5–5v; 1663, IV, 224; 1968, 17.

[35] *numerus quadratorum [...] semper componitur ex tribus aestimationibus iunctis simul.* [The number of squares [...] is always composed of the three solutions taken together.] Cardano 1545, 39v; 1663, IV, 259; 1968. 134.

[36] *quo naturae nõ licet.* [Which in nature is not allowed.] Cardano 1545, 3v; 1663, IV, 222; 1968, 9.

or cubes were regarded as meaningless. As we have seen, this does not imply either unwillingness or inability to handle negative quantities, only the perception that there was not much point in doing so.

In Chapter 1, all Cardano's examples yielded integer roots. In Chapter 4, he spoke briefly about roots of a more complicated kind. For quadratic equations, for example, he claimed that roots could take the form $l \pm \sqrt{m}$ or $\sqrt{l} \pm m$ (but not $\sqrt{l} \pm \sqrt{m}$, which suggests that he regarded the coefficients of the equation as integers or fractions only).[37] Roots of cubic equations, meanwhile, could be of the form $\sqrt[3]{p} \pm \sqrt[3]{q}$ or even $\sqrt[3]{p} \pm n \pm \sqrt[3]{q}$. In his illustrative example in Chapter 4, the quantities p, q, n, are integers (he suggested $\sqrt[3]{16} \pm 2 + \sqrt[3]{2}$), but he also remarked that n is one-third of the coefficient of the square term, and therefore it could obviously be a fraction, while many other examples elsewhere in the book show that p and q could be of the form $l \pm \sqrt{m}$ or $\sqrt{l} \pm m$. In Chapter 6 he actually experimented with substituting roots of the form $l + \sqrt{m}$ into an equation of the form $x^3 + d = cx^2$ and observed that he could equate rational and irrational parts separately. Thus he discovered that a root of $x^3 + 3x^2 = 14x + 20$ is $1 + \sqrt{5}$, for example.[38]

He did not discuss the structure of the roots of quartic equations, though from his examples it was clear that such roots, derived as they were from solving first a cubic equation and then a pair of quadratics, contained square roots in the outer layer with cube roots and possibly further square roots nested inside those. His only comment on equations of higher degree was that a square, cube, or fifth root taken alone could satisfy only an equation of the simplest form, that is, a power equal to a number, but not any compound equation; and likewise a simple (non-compound) equation could not be satisfied by a sum of such roots.[39]

Finally, in Chapter 37, Cardano returned to the concept of negative roots by means of examples that require one to find money owed or lacking. Here too he raised the topic of negative squares, with the problem of finding two numbers whose sum is 10 and whose product is 40. This gives rise to the equation $x^2 + 40 = 10x$, and the rules for solution yield $5 + \sqrt{-15}$ and $5 - \sqrt{-15}$. Cardano satisfied himself that these numbers fit the requirements but was not altogether happy with his own geometric 'demonstration'. He complained that it required the comparison of a square with a line, which geometrically speaking is dimensionally inconsistent, but the true problem was perhaps that the demonstration involved a negative area, which in his world was meaningless.

---

[37]The solution of $x^2 + bx + c = 0$ is $x = -\frac{b}{2} \pm \sqrt{\left(\frac{b}{2}\right)^2 + c}$ so we can conclude from Cardano's assertions that he took both b and c to be rational.

[38]*si dixero cubus & 3q̃drata, aequalia sunt 14 rebus, & 20 numero, & ponatur quantitas quaedã intellecta, aestimatio rei, cuius prima pars sit numerus, secunda vero quantitas, alia pars irrationalis. Et fit gratia exempli, hic 1 p̃;R5.* [If I say that a cube and 3 squares are equal to 14 things and to 20 in numbers, and there is put a certain understood quantity, then in the estimated thing, in which the first part is a number, that is a true quantity, the other part will be irrational. And it becomes here, for example, 1 plus the root of 5.] Cardano 1545, 15v; 1663, IV, 235; 1968, 50.

[39]*& sicut hae simplices composites capitulis convenire nequeunt, sic nec ullum compositũ ex pluribus radicibus incommensurabilis capitulo simplici potest convenire.* [and just as these simple roots cannot satisfy any compound equation, so no sum of incommensurable radicals can satisfy a simple equation.] Cardano, 1545, 9; 1663, IV, 229; 1968, 32.

Taking the *Ars magna* as a whole it is clear that it contains much more than just a set of rules for solving equations. For convenience we may summarize Cardano's major achievements as follows.

1. A general rule for solving cubic equations, with particular rules for the irreducible case where the general rule appears not to work.
2. An algorithm, demonstrated by worked examples, that can be applied to solving any quartic equation.
3. An understanding that roots of equations can be positive or negative, and in the quadratic case a hint that they could even be imaginary.
4. An investigation of the number of real roots, and whether they are positive and negative, of any cubic equation.
5. An understanding that roots of quadratic equations (with rational coefficients) are sums of rationals and square roots, and that sometimes the square root might be of a negative quantity; and that the roots of cubic equations can be combinations of rationals and cube roots.
6. The observation that a substitution of a number of the form $l \pm \sqrt{m}$ into a polynomial equation gives rise to two separate equalities, in rational and irrational quantities respectively.
7. The insight that equations can be transformed from one kind to another by simple substitutions. Those that Cardano used were of the form

   (a) $x \to -x$,
   (b) $x \to k/x$,
   (c) $x \to x \pm k$.

8. A special interest in three-term equations of the of the form $x^n + q = px^m$, with rules for their solution.
9. A rudimentary attempt to find an approximate numerical solution when the exact solution is not easily found, the first known published discussion of this problem by a European writer.

Because Cardano had no general notation for coefficients of equations, all his rules and insights were demonstrated by means of specific examples, though it was usually quite clear that he had general applications in mind. The *Ars magna* is thus a collection of rules, special cases, and techniques rather than an attempt at a theory in the sense we would now understand it. Nevertheless, it went very much further than any previous textbook, and many of the features outlined above were to recur repeatedly in later treatments.

Bombelli's *Algebra*, 1572

The public dispute between Tartaglia and Cardano following the publication of the *Ars magna* meant that the book rapidly became well known, if not well understood, in the university towns of northern Italy. In particular it came to the attention of Rafael Bombelli in Bologna. During the 1550s Bombelli was employed in draining the lakes and marshes of the Chiana valley (between Siena and Arezzo) but he turned to the study of algebra when the project was temporarily suspended sometime after 1555, and wrote his own *Algebra*, in Italian, between 1557 and 1560. It consisted of five books, of which Books I to III were published in 1572, the year of Bombelli's death. Books IV and V, which explore the relationship between algebra and geometry, remained unpublished until 1923.

Bombelli greatly admired Cardano's work but found his exposition somewhat obscure.[40] Much of his own *Algebra* is essentially a re-writing of the *Ars magna*, in a much clearer and better organized style. One noticeable improvement in Bombelli's text compared with Cardano's is a more useful notation for powers: $\overset{1}{\smile}$ for 'things', $\overset{2}{\smile}$ for squares, $\overset{3}{\smile}$ for cubes, and so on. Thus the equation we would now write as $x^3 = 6x + 20$ appears as 1 $\overset{1}{\smile}$ *a*. 6 $\overset{1}{\smile}$ *p*. 20, where '*a*' stands for *aggualisi* (equals) and '*p*' for *piu* 'plus'. As in the discussion of Cardano's work, we will here fall back on the equivalent modern notation.

Book I is 195 pages long, and consists entirely of a treatment of powers, roots, binomes and residuals (quantities of the form $l \pm \sqrt{m}$ where l and m are integers). The final 20 pages teach the handling of what Bombelli called *piu di meno* $(+\sqrt{-1})$ and *meno di meno* $(-\sqrt{-1})$, abbreviated to *p.di m* and *m.di m*, respectively. He also wrote, for instance, *p.di m*.2 for $+2\sqrt{-1}$. By manipulating such quantities arithmetically, Bombelli was able to show by the following calculation that $(-\sqrt{-1}-1)^3 = 2-2\sqrt{-1}$.

m.di m.1.*m*.1.	$-1\sqrt{-1}-1$
m.di m.1.*m*.1.	$-1\sqrt{-1}-1$
m.1.*p.di m*.1.*p.di m*1.*p*.1.	$-1+1\sqrt{-1}+1\sqrt{-1}+1$
p.di m.2.	$+2\sqrt{-1}$
m.di m.1.*m*.1	$-1\sqrt{-1}-1$
cubato 2.*m.di m*.2	*the cube is* $2-2\sqrt{-1}$

A similar calculation for $(-\sqrt{-1}+1)^3$ shows that $(-\sqrt{-1}-1)^3 + (-\sqrt{-1}+1)^3$, which at first sight appears to be an 'impossible' number, is in fact equal to 4, because the 'imaginary' parts *m.di m*.2 $(-2\sqrt{-1})$ and *p.di m*.2 $(+2\sqrt{-1})$ cancel each other out.

Bombelli's treatment of quadratic, cubic, and quartic equations is in Book II. Like Cardano and other contemporary writers he wrote equations as relationships between positive terms, and dealt with each possible case separately: 3 cases for quadratics, 13 cases for cubic, and 43 cases for quartics. His treatment of quadratics was standard. For cubics, right from the beginning, he taught the transformations by which equations of one type could be changed to equations of another. These were exactly those given by Cardano: replace x by k/x or by $x \pm k$ for some suitable value of k. Bombelli's

[40] *... in vero alcuno non è stato, che nel secreto della cosa sia penetrato, oltre che il Cardano Melanese nella sua arte magna, oue di questa scientia assai disse, ma nel dire sù oscuro;* [In truth there is no-one who has penetrated so far into the secrets of the unknown quantity (*cosa*) as Cardano of Milan in his *Ars magna*, where he has said much on this science, but has said it obscurely;] Bombelli 1572, 'Agli lettori' ('To the reader').

exposition was not given in such general terms, of course, but through well chosen worked examples for each case, together with geometric demonstrations for a few of them. He also discussed the use of the quantity $\frac{4}{27}c^3 - d^2$ for determining the number and nature of the roots of cubics of the type $x^3 + d = cx$. For quartics he taught the method devised by Ferrari and Cardano, but with many more examples, systematically arranged by case ('fourth power and root', 'fourth power and cube', 'fourth power, square, and root', 'fourth power, cube, and root', and so on.

Bombelli's treatment was followed closely a few years later by Simon Stevin in his *L'arithmetique ... aussi l'algebre*, published in 1585 when Stevin was living in Leiden. Stevin, an engineer himself, greatly admired Bombelli, whom he described as 'a great arithmetician of our time' (*grand Arithmetician de nostre temps*).[41] In particular, he took up Bombelli's circle notation for powers, which, together with his use of + and − symbols, makes his text much easier on the eye for a modern reader than most algebraic writings of the sixteenth century. Less easy to understand are Stevin's idiosyncratic descriptions of equations in terms of proportions. Here, for example, is his approach to the cubic equation that we would write as $x^3 = 6x + 40$ nowadays.[42]

> Suppose there are three terms in the problem as follows: the first is 1③, the second is 6①+40 the third is 1①. One must find their fourth proportional term.

Stevin then calculated, using Cardano's rule, that the required root is 4, and set out the elements of the problem as a table of proportionals:

1③	6① + 40	1①	4
64	64	4	4

That is, in modern notation,

$$x^3 : 6x + 40 = x : 4.$$

In most other ways, Stevin's exposition was very clear. He began with the usual rules for simplifying and rearranging equations, and then worked systematically through the various cases of quadratic, cubic, and quartic, using the same rules and transformations as Cardano and Bombelli, and offering the full details of each calculation.

Viète's *Tractatus duo*, 1615

François Viète was born in western France in 1540, and studied law at the University of Poitiers. During his twenties he acted as tutor to Catherine of Parthenay, daughter of a local aristocratic family, and his lectures to her on geography and astronomy were later printed as *Principes de cosmographie* (1637). During this period he also worked on plane and spherical trigonometry but only part of it was ever published, as his *Canon*

[41] Stevin 1585, 269; 1958, 586.

[42] *Soyent donnez trois termes selon le probleme tels: le premier* 1③*, le second* 6① + 40 *le troisiesme* 1①*. Il faut trouver leur quatriesme terme proportionel.* Stevin 1585, 305–306; 1958, 615–616.

mathematicus (1579). Viète moved to Paris in 1570, and thereafter became a counselor to the *Parlements* (courts of justice) of Paris and Brittany, and a royal privy counselor in 1580. From 1584 to 1589 he was exiled from the court for complicated political reasons, and once again turned to mathematics. His ideas on algebra were almost certainly worked out during these years of comparative leisure. Afterwards he returned to political office and lived in Tours until 1594 but then mostly in Paris until his death in 1603.

From 1588, or possibly earlier, Viète worked as a cryptanalyst, and some have seen connections between this and his innovations in algebra.[43] Such speculations, however, can easily become rather fanciful. The fact that Viète sought out general methods both in code-breaking and in algebra may be seen as the mark of an intelligent mind rather than of an intrinsic connection between the two activities. His decipherments were based essentially on frequency analysis, a very different technique from any he used in algebra. His recognition that apparently random letters can represent comprehensible text may have influenced his view that the symbols in an equation can represent either arithmetic or geometric quantities according to context, but we cannot be sure of it. Certainly, as Pesic has pointed out (1997b), the primary need to distinguish between vowels and consonants in code-breaking could well have led to Viète's use of the same distinction in algebra, where he used the vowels A, E, ... for unknown quantities, and consonants B, C, D, ... for known or given quantities. The precise form of his symbolism, however, is less important than its existence, which seems to me to arise quite naturally from certain mathematical requirements, discussed below. It is true that Viète's notation was crucial in allowing discussion of equations to move beyond representative numerical cases to general literal forms. When writing particular equations with numerical coefficients, however, he fell back on the older cossist notation in which C represents a cube, Q a square, and R or N either a root or an unknown number. In either system he expressed operations and connections verbally (apart from the symbols $+$ and $-$) so that his writing still has much of the appearance of older verbal texts. When Viète started writing on algebra in the early 1590s he may not have known about the notational advances of Bombelli or Stevin for writing powers (though he did by 1595, see Chapter 3, note 19); if he did, he ignored them, falling back instead on expressions like *A-quadratus* and *A-cubus*, even though these offered no way of writing a general power of unknown dimension.

Tantalising and unanswered questions remain about other influences on Viète's mathematics. We do not know which writers on algebra he had read, but almost certainly Cardano was one of them since much of Viète's later work on equations followed and extended what was in the *Ars magna*. He was certainly thoroughly familiar with the classical geometry collected and expounded in Pappus' *Synagoge*. Of particular importance to Viète was Pappus' discussion of 'analysis' and 'synthesis' in Book VIII. 'Analysis', according to Pappus, was a procedure in which one assumed that a theorem was true, or a problem solved, and then worked backwards to discover the foundations on which the theorem or problem rested; from there one could reconstruct

[43] Pesic 1997a, 1997b.

a proof or solution by working in the opposite direction, that is, by 'synthesis'. A common complaint of seventeenth-century mathematicians was that classical writers had presented only their final results, hiding the process of analysis by which they were thought to have discovered them.

Viète's new vision of algebra was published in a series of short privately distributed treatises from 1591 onwards, the first of which was the *Isagoge in artem analyticem* (*Introduction to the analytic art*) (1591). Almost all writers on algebra before Viète had used geometric squares, rectangles, and cubes to represent or justify algebraic manipulations. Viète, however, began to understand the power of the relationship in the other direction. He saw more clearly than any previous writer that the unknown quantities in algebraic equations could correspond either to numbers or to geometric magnitudes, and that one could therefore move smoothly backwards and forwards between geometric constructions and equations. In recognizing algebra as a tool for opening up geometric problems, he came to identify it with the method of 'analysis' that had supposedly been used but hidden by the ancients. In Viète's hands, algebra was transformed from the simple *regula cosa* (rule of 'things') of earlier writers to a sophisticated new technique, the 'analytic art'.

Viète's ideas were set out in condensed form in the *Isagoge*, but were developed at much greater length in the subsequent treatises, which together made up his *Opus restitutae mathematicae analyseos seu algebra nova* (*The work of restoration of mathematical analysis, or the new algebra*). Some of these treatises examined in detail the relationship between algebraic equations and geometric constructions, particularly the *Effectionum geometricarum canonica recensio* (1593) and *Supplementum geometriae* (1593). Another, the *Zetetica libri quinque* (1591 or 1593) took up a number of problems from Diophantus and showed how they too could be represented by equations.

Two further treatises, *De numerosa potestatum ad exegesin resolutione* (*On the numerical resolution of powers*) (1600) and *De recognitione et emendatione aequationum tractatus duo* (*Two treatises on understanding and changing equations*) (1615), dealt specifically with understanding and solving equations. In a list Viète gave at the end of the *Isagoge* of the ten treatises he intended to publish, these came fourth and fifth, respectively, but the eventual order of publication of his work was more haphazard. *De resolutione* was published in 1600 but the *Tractatus duo* came out only in 1615, edited by Alexander Anderson twelve years after Viète's death. The first part of the *Tractatus duo*, 'De recognitione', was almost certainly completed along with several other treatises in the early 1590s; the second part, 'De emendatione', was possibly added later. Both parts offer a theoretical treatment of equations. From a historical point of view, they fall naturally alongside the treatises of Cardano and Bombelli, and are therefore described in this chapter. *De numerosa potestatum resolutione*, on the other hand, will be discussed in the next.

Like all of Viète's writings, the *Tractatus duo* is dense and difficult. Viète frequently borrowed or invented Greek terms to describe special cases and techniques but such words carry little or no meaning for a modern reader. Further, Viète's conceptual framework was embedded in Greek concepts of ratio; almost all his writing is couched in the language of proportion, a mode of description that was all-pervasive in the early modern period but which has all but disappeared from modern mathematics. Viète was perhaps more keen to emphasize Greek ideas than to acknowledge the Islamic influences at work in Renaissance algebra, and yet in some senses his greatest achievement was his marrying of the two by applying the techniques of Islamic algebra to Greek geometry. Anyone who wanted to apply algebra to geometry, however, must have a thorough understanding of equations, and this was what the *Tractatus duo* was meant to provide.

At the beginning of the *Tractatus duo* Viète stated that his concern was to explain the structure (*constitutione*) of equations as an aid to solving them. In a glorious mixture of metaphors he asked: 'Surely no Analyst will start out without understanding the structure of a proposed equation, so that he can avoid the rocks and reefs? And like an expert anatomist, turn it around, hold it down, raise it up, and at all times operate safely?'[44]

The structures that Viète had in mind were all described in terms of proportions. His first example is the equation that he wrote as *A quad + B in A, aequatur Z quad*. Recall that for Viète, A was an unknown quantity, but B and Z were supposed known. For convenience we can write his equation in modern notation as

$$A^2 + BA = Z^2.$$

For Viète, such an equation was equivalent to a statement about three quantities in geometric proportion. He regarded A as the first and smallest of them, and B as the difference between the first and third, so that the third and largest is $A + B$; finally the quantity Z is the middle quantity, or the geometric mean of the other two.[45] We may thus write the three quantities in increasing order of size as

$$A,\ Z,\ A + B$$

from which it follows immediately, as required, that

$$A(A + B) = Z^2. \tag{7}$$

If A is taken to be the largest quantity instead of the smallest, the quantities will be

$$A - B,\ Z,\ A$$

and the equation connecting them will be

$$A(A - B) = Z^2 \tag{8}$$

which for Viète, as for his predecessors, was a different kind of quadratic equation from (7). The third and last kind of quadratic equation arises when Z is the geometric mean,

[44] *Ecquid vero aequationis, quae proposita[e] est, agnita constitutione non tentabit Analysta, quo saxa & scopulos refugiat? num gnarus Anatomices invertet, deprimet, attollet, & undique operabitur secure?* Viète 1646, 84; 1983, 160.

[45] *Sunt tres proportionales radices, quarum media est Z, differentia vero extremarum B; & fit A minor extrema.* [There are three proportional quantities, of which the mean is Z, and the difference between the extremes B; and A is the smaller extreme.] Viète 1646, 85; 1983, 161.

but B is the sum of the first and third quantities.[46] In this case A can be either the first or the third quantity, that is, the three proportionals are either

$$A,\ Z,\ B - A$$

or

$$B - A,\ Z,\ A.$$

Either arrangement gives the equation

$$A(B - A) = Z^2, \tag{9}$$

which has two positive roots (because if A is a root then so is $B - A$). Thus Viète could describe all three standard cases of quadratic equation as relationships between three proportional quantities.

For cubic equations, Viète needed four proportional quantities. For him, the equation

$$A^3 + B^2 A = B^2 Z$$

corresponded to the statement:[47]

> There are four continued proportionals, of which the first, whether it is the greater or smaller of the extremes, is B, and the sum of the second and the fourth is Z, and A is the second.

To see how this works let us borrow modern notation and call the four proportional quantities a, ar, ar^2, ar^3. Viète called the first of these B (which may be either the greatest or the smallest, depending on whether the quantities are increasing or decreasing), that is, $B = a$. Next he stated that Z is the sum of the second and the fourth, that is, $Z = ar + ar^3$. Finally he claimed that A is the second, that is, $A = ar$. The first, second, and fourth quantities, namely,

$$a,\ ar,\ ar^3$$

can therefore be written in Viète's notation as

$$B,\ A,\ Z - A.$$

Now it is clearly always true that

$$(ar)^3 = a^2 \times ar^3,$$

[46] *sunt tres proportionales, quarum media est Z, aggregatum B; & fit A minor* [*major*], *minorve extrema.* [There are three proportional quantities, of which the mean is Z, and the sum B; and A is the greater or smaller extreme.] Viète 1646, 86; 1983, 163.

[47] *sunt quatuor continue proportionales, quarum prima majorminorve inter extremas est B, aggregatum vero secundae & quartae est Z, & fit A secunda.* Viète 1646, 86; 1983, 164.

that is, that the cube of the second quantity is the product of the fourth and the square of the first. In Viète's notation, this gives

$$A^3 = B^2(Z - A),$$

which transposes to the required equation.

As an example, Viète observed that in the equation $1C + 64N$ *aequari* 2496 (or $x^3 + 64x = 2496$), we have $B^2 = 64$ and $B^2Z = 2496$, giving us $B = 8$ and $Z = 39$. Viète then claimed that the four proportional quantities are 8, 12, 18, 27, and the root of the equation is therefore 12, the second of them. He did not explain, however, how to discover that the last three numbers are 12, 18, 27; further, any attempt to find them, where they cannot be seen by inspection, leads only to another cubic equation. In other words the proportionality relationship that underlies the equation explains how the root A is related to the known quantities B and Z, but does not offer a way finding it. Similar ideas of proportion almost certainly lay behind Cardano's instructions for finding the roots of three-term equations: in these examples too, numbers had be found that fitted proportional relationships between the coefficients, but the only technique of discovering them was by inspection.

In Chapter 7 of the *Tractatus duo*, Viète moved on to the transformation of equations, using substitutions either of the form $E = A \pm B$ or else $E = BA$, or $E^2 = BA$, and so on, giving numerous examples in this and the next six chapters.

One of the more interesting and significant sections of 'De recognitione' comes towards the end, in Chapter 16, which is entitled 'De syncrisi' ('On syncrisis' or 'On comparison'). As we have seen, Cardano was particularly interested in three-term equations, and so was Viète. The forms $x^n \pm px^m = q$ (with $n > m$ and $p > 0, q > 0$) have just one positive root, while the form $px^m - x^n = q$ may have two, depending on the relative sizes of p and q. These facts had long been known for quadratic equations ($n = 2, m = 1$) and since Cardano also for cubic equations ($n = 3, m = 1$ or 2). Viète is likely to have discovered them also for higher degree three-term equations ($n \geq 4$) from his experience of equation-solving (see pages 32–33).

Moving closer to Viète's notation, suppose we have an equation $ba^m - a^n = z$ with two positive roots.[48] Viète denoted the roots by A and E, and took A to be greater than E. He then argued that

$$bA^m - A^n = z$$

and

$$bE^m - E^n = z.$$

Hence we can write

$$bA^m - A^n = bE^m - E^n$$

[48] Viète 1646, 105–107; 1983, 208–209. There are two points to note here. (i) For the purposes of matching algebra to geometry Viète assumed throughout his work that equations were dimensionally homogeneous. Here, therefore, b must be assumed to be of dimension $n - m$ and z of dimension n. (ii) For any $n \geq 3$ the equation $ba^m - a^n = z$ can have one, two, or three real roots, but no more. For Viète's argument to work it must be assumed that there are at least two real roots. Viète took them to be positive, but his argument is valid for any combinations of sign.

from which we have

$$b = \frac{A^n - E^n}{A^m - E^m}$$

and

$$z = \frac{A^n E^m - E^n A^m}{A^m - E^m}.$$

In other words, the coefficient b and the 'homogene' z (the term free of a), can both be expressed in terms of the two roots A and E. Viète gave the name *syncrisis* (comparison) to this process of comparing the equation in A with the equation in E. By applying *syncrisis* to a quadratic equation of the form $ba - a^2 = z$ Viète showed that $b = A + E$ and $z = AE$. For a cubic of the form $ba - a^3 = z$ he found the less obvious result that $b = A^2 + E^2 + AE$ and $z = A^2E + E^2A$. Viète was able to carry out a similar procedure for other cases of three-term equations but it was the type described here, with two positive roots, that was to be important later.[49]

The second part of Viète's treatise, entitled 'De emendatione aequationum', deals with the 'emendation' or transformation of equations. Here Viète again worked with the transformation $E = A \pm B$ but now with the specific objective of removing the second term from either a quadratic or a cubic.[50] He also taught the transformation that had first led Cardano into the mysteries of cubics, namely $E = Z/A$. Like Cardano, Viète applied it to cubics of the form 'cube, root, number' but with a slightly different purpose in mind: Viète's aim was not to change a square term to a linear term (or vice versa) but to change a negative term to a positive.[51] Thus, for instance, by replacing $1N$ by $40/(1N)$, he could change $1C - 96N$ *aequari* 40, which is 'negatively affected' to $1C + 96Q$ *aequari* 1600, which is 'positively affected'.

He also explored Cardano's technique of reducing the degree of an equation by suitable division:[52] recall from above how Cardano reduced $x^3 = 5x + 2$ to $x^2 = 2x + 1$ by adding 8 to each side and dividing by $x + 2$. Viète with his love of Greek terms called this method *anastrophe* (turning back). It only works, however, when it is possible to adjust the equation in such a way that a suitable divisor is easily spotted. Viète used it for reducing cubics to quadratics, or quintics to quartics.

In Chapter 6 of 'De emendatione', Viète moved on to quartics, which for some reason he tackled before cubics. His method was exactly that developed by Ferrari and Cardano, whereby a quartic is reduced by means of a cubic to a product of quadratics. Where Cardano had given just seven examples, Viète gave twenty, covering cases such as $A^4 + BA = Z$ and $BA - GA^2 - A^4 = Z$. In each case he gave the general form of the intermediate cubic.[53]

[49]The repercussions of this method for practical equation-solving are discussed on pages 32–33. Beyond that, in the 1620s, Fermat took up the method in the course of his work on maxima and minima, noting that at a maximum (or minimum) two previously distinct roots will coincide. Viète's method of *syncrisis* gave Fermat important information about the conditions under which this would happen. For a full discussion of Fermat's insights see Mahoney 1994, 147–157. Note that the equations cited by Mahoney on pages 153 and 154, $bx - x^2 = z$ and $bx^2 - x^3 = z$, are both of the form discussed above.

[50]Viète 1646, 127–132; 1983, 240–246.

[51]Viète 1646, 132–134; 1983, 246–250.

[52]Viète 1646, 134–138; 1983, 250–260.

[53]Viète 1646, 140–148; 1983, 266–286.

Afterwards Viète turned to cubics, but developed a method different from Cardano's, though it leads to the same result. For cubics of the form $A^3 + 3BA = 2Z$, Viète used the substitution

$$A = \frac{B - E^2}{E},$$

which leads to the equation

$$(E^3)^2 + 2ZE^3 = B^3.$$

This is a quadratic in E^3 and hence solvable, and once E is found it can be substituted back to find A. Similar substitutions with appropriate changes of sign can be used for other cubics lacking a square term.[54]

Viète's treatise ends with a list of special cases that may be solved by specific techniques or by inspection.[55] It is easily seen, for example, that

$$A^3 - 3B^2A = 2B^3$$

is satisfied by $A = 2B$, or that

$$BA^2 + D^2A - A^3 = D^2B$$

is satisfied either by $A = B$ or by $A = D$.

In a final short section Viète dealt with equations that have all their roots positive. The first that he gave is

$$(B + D)A - A^2 = BD,$$

with roots B and D, and the last is

$$\begin{aligned} A^5 &+ (-B - D - G - H - K)A^4 \\ &+ (+BD + BG + BH + BK + DG + DH + DK + GH + GK + HK)A^3 \\ &+ (-BDG - BDH - BDK - BGH - BGK \\ &\quad - BHK - DGH - DGK - DHK - GHK)A^2 \\ &+ (+BDGH + BDGK + BDHK + BGHK + DGHK)A = BDGHK, \end{aligned}$$

which is satisfied by putting A equal to any of B, D, G, H, K. Viète called the reasoning out of this observation the crowning achievement of his treatise.[56] He did not give his reasoning but almost certainly it was based on his well tried method of *syncrisis*. For a cubic equation with three positive roots the method would work like this. Suppose the equation $A^3 - RA^2 + SA = Z$ has roots B, D, G. Viète would

[54] Viète 1646, 149–150; 1983, 286–289.

[55] Viète 1646, 152–158; 1983, 293–310.

[56] *Atque haec elegans & sepulchrae speculationis sylloge, tractatui alioquin effuso, finem aliquem & Coronida tandem imponito.* [Indeed the elegant reasoning out of this beautiful observation, which I have otherwise treated extensively, I place here as the end and in some ways the crown.] Viète 1646, 158; 1983, 310.

have assumed $B < D < G$, but the ordering is not essential to the argument as long as the roots are distinct. Then he could say that

$$B^3 - RB^2 + SB = Z, \tag{10}$$

$$D^3 - RD^2 + SD = Z, \tag{11}$$

$$G^3 - RG^2 + SG = Z. \tag{12}$$

Thus, from (10) and (11),

$$B^3 - RB^2 + SB = D^3 - RD^2 + SD$$

or

$$(D^3 - B^3) - R(D^2 - B^2) + S(D - B) = 0.$$

Dividing by $(D - B)$ gives

$$(B^2 + BD + D^2) - R(B + D) + S = 0. \tag{13}$$

By a similar deduction from (11) and (12) we also have

$$(D^2 + DG + G^2) - R(D + G) + S = 0. \tag{14}$$

Subtracting (14) from (13) gives

$$(B^2 + BD - DG - G^2) - R(B - G) = 0,$$

and dividing by $(B - G)$ gives

$$R = B + D + G.$$

Substitute this back into (10) and (11) to get (after a little simplification)

$$-B^2D - B^2G + SB = Z \tag{15}$$

and

$$-BD^2 - D^2G + SD = Z. \tag{16}$$

Now subtract (15) from (16) and divide by $(D - B)$ to get

$$S = BD + DG + BG.$$

Finally, put R and S back into (10) to get

$$Z = BDG.$$

It would not have been difficult for Viète to extend this argument to an equation with four or even five roots. Once he had the coefficients he could easily check that such equations really were satisfied by the values $A = B$, $A = D$, $A = G$, and so on.

How may we summarize Viète's achievements? Cardano, Bombelli, and Stevin had all given general treatments of equations, but all of them had done so through worked examples of particular cases. Viète instead wrote each kind of equation in general notation, using the letters $B, D, F, G, \ldots$ for coefficients (avoiding C probably because he used it elsewhere to stand for cubes). In this way he was able not only to write down general rules for transforming equations, but also to write in general form what the results of particular substitutions or transformations would be. Thus Viète's treatment appears to be much closer than Cardano's to what we expect a general theory of equations to look like. Most of Viète's methods and results, however, were extensions or generalizations of Cardano's, the main exception being his method of *syncrisis*.

Viète's notational advances, however, were just one aspect of what, in my view, was his most outstanding contribution to mathematics, the reversal of the older perception of algebra as dependent on or justified by geometry. Viète gave algebra a startling new priority as a tool for investigating and analysing the problems and theorems of classical geometry. Even the hitherto intractable difficulties of doubling the cube or trisecting an angle were now, in his opinion, amenable to algebraic treatment: Viète could show that the trisection problem, for instance, reduces to a cubic equation. This new vision of the scope and power of algebra forced him to examine the nature and construction of equations much more carefully than any of his predecessors had done. Thus, in his *Effectionum geometricarum* and *Supplementum geometriae* (both published in 1593) Viète demonstrated geometric constructions that correspond or give rise to equations of second, third, or fourth degree.

Even for equations of higher degree, where direct geometric representations fail, Viète's grasp of the theory of proportions enabled him to analyse three-term equations of the form $px^m \pm q = x^n$. Beyond that, however, ideas about proportion ceased to be helpful, and indeed possibly blocked other and more fruitful approaches. The next important developments in the theory of equations were to be influenced not by the publication of the *Tractatus duo* but by Viète's much more practical book, *De numerosa potestatum resolutione*, with which we will begin the next chapter.

Chapter 2
From Viète to Descartes

Both chronologically and mathematically Viète stood at the cusp between the sixteenth century and the seventeenth. He began publishing his most important work in 1591 but died in 1603 just as the new century was beginning. In this present book, he belongs both to the first chapter, where his work stands as the culmination of the sixteenth-century theory of equations, but also just as certainly to this one, where we examine his influence on the mathematics of the early seventeenth century. Viète's formation and motivation were rooted in the classical texts of the Renaissance, yet possibly he more than anyone else propelled mathematics into a new and very different era. His analytic art created a fusion of geometry and algebra that was to have a profound influence in the years that followed. Less widely recognized has been his work on the numerical solution of equations, which in the hands of Thomas Harriot was to lead away from the understanding of equations as relationships between proportional quantities, and into completely new ideas about the structure of equations.

The second part of this chapter takes us fully into the seventeenth century with the work of Girard and Descartes. The brief comments on equations made by Descartes were to become the foundation of much further work. Descartes stood so large in seventeenth-century mathematics that his predecessors slipped into the shadows, and Descartes was content to leave them there; questions about the influence of Viète or Harriot on Descartes therefore remain to this day tantalisingly unanswered, and the reader must draw his or her own conclusions on the matter.

Viète's *De numerosa potestatum resolutione*, 1600

Viète's *De numerosa potestatum ad exegesin resolutione* (*Towards showing the numerical solution of equations)*, published in Paris in 1600, was quite different in character from the *Tractatus duo*, discussed in Chapter 1. *De resolutione* was not a theoretical text but a practical one, the first of its kind, which taught how to find roots of polynomial equations by a method of successive approximation. For Viète, such a method was essential to his vision of leaving no problem unsolved.[1] The intractability of the classical problems of doubling a cube or trisecting an angle, for instance, lay not in arriving at the right equations but in the difficulty of solving them.

Viète demonstrated his method first for simple powers (*potestates purae*): squares, cubes, fourth, fifth, and sixth powers. It is explained here by an example, borrowed from Viète but simplified a little by the use of modern notation.[2] Suppose we wish to find the cube root of 157 464. By inspection, we can see that the root must lie between

[1]Viète ended his *Isagoge* with words that represented both his hopes and his idiosyncratic use of Latin: *nullum non problema solvere* [to leave no problem unsolved] Viète 1646, 12; 1983, 32.

[2]Viète 1646, 166–168; 1983, 317–319.

50 and 60 and so its first digit, the tens digit, must be 5. Suppose that the root is in fact $50 + y$, so that $(50 + y)^3 = 157\,464$. Thus we have

$$7500y + 150y^2 + y^3 = 32\,464. \tag{1}$$

Viète now calculated an estimate for y by dividing 32 464 by $7500 + 150 = 7650$. In other words, he neglected y^3 and replaced y^2 by y. This is a rough and ready method, to be sure, but it suggests that y must be close to 4. In fact $y = 4$ satisfies equation (1) exactly since $30\,000 + 2400 + 64 = 32\,464$. Thus the required cube root is 54. If further digits had been needed, they could have been found, as Viète indicated, by adjoining zeros (three at a time) to the original number.

Thus the method works by eliciting successive digits of the root in turn. In Viète's treatise the calculations are written in the following tabular layout:

Calculation for the first digit.

1	5	7	4	6	4
1	2	5			
	3	2	4	6	4

Calculation for the second digit.

3	2	4	6	4
	7	5		
		1	5	
	7	6	5	
3	0	0		
	2	4	0	
		6	4	
3	2	4	6	4

It is essential to keep the entries correctly aligned and Viète gave careful instructions for doing so. He also annotated each row to explain where it came from. He used none of the symbolic notation that appears in the *Tractatus duo*; thus equation (1), written above as $7500y + 150y^2 + y^3 = 32\,464$, was described by Viète verbally and in geometric terminology as follows:[3]

> The total number remaining, 32 464, consists of the solid formed by the square of the side of the second and three times the first $[y^2 \times 3 \times 50]$, plus the solid formed by three times the square of the first and the side of the second $[\,3 \times 50^2 \times y]$, to be found, plus the cube of the second $[y^3]$.

Viète next turned to 'affected powers' (*potestates adfectae*), where the leading power is 'affected' by the addition or subtraction of lower powers. He did not explain

[3] *Unde totius numerus residuus 32,464 constans solido sub lateris secundi quadrato & triplo primi, plus solido sub triplo quadrato primi & latere secundo inveniendo, plus cubo secundi.* Viète 1646, 167; 1983, 318.

his method but, as with his method for a cube root, his calculations reveal his procedure. Once again the method is illustrated here with one of his own examples,[4] the equation Viète wrote as $1C + 95{,}400N$ *aequari* 1,819,459, which for convenience we will write as $x^3 + 95\,400x = 1\,819\,459$. This time inspection shows that the root lies between 10 and 20, and so the first digit, the tens digit, is 1. That is, we take 10 as a first approximation. Now suppose that $10 + y$ is a second and better approximation. Expanding $(10 + y)^3 + 95\,400(10 + y)$, we find that y must satisfy

$$y^3 + 30y^2 + 300y + 95\,400y = 864\,459. \qquad (2)$$

As before, neglecting y^3 and replacing y^2 by y, Viète divided 864 459 by $95\,400 + 300 + 30 = 95730$, which suggests that the next digit of the solution is close to 9. It is easily checked that 9 in fact satisfies equation (2) exactly. Thus 19 is a root of the original equation.

Viète demonstrated his method on the following 'positively affected' equations, all of whose solutions are two- or three-digit integers:

$$\begin{aligned}
&x^2 + 7x = 60\,750,\\
&954x + x^2 = 18\,847,\\
&x^3 + 30x = 14\,356\,197,\\
&x^3 + 95\,400x = 1\,819\,459,\\
&x^3 + 30x^2 = 86\,220\,288,\\
&10\,000x^2 + x^3 = 5\,773\,824,\\
&x^4 + 200x^2 = 446\,976,\\
&x^4 + 200x^2 + 100x = 449\,376,\\
&x^6 + 6000x = 191\,246\,976;
\end{aligned}$$

and on the following, which are 'negatively affected':

$$\begin{aligned}
&x^2 - 7x = 60\,750,\\
&x^2 - 240x = 484,\\
&x^3 - 10x = 13\,584,\\
&x^3 - 116\,620x = 352\,947,\\
&\dots\\
&\dots\\
&x^5 - 5x^3 + 500x = 7\,905\,504;
\end{aligned}$$

and on these, which he called 'avulsed powers' (*potestates avulsae*), literally powers that are 'torn away':

$$\begin{aligned}
&370x - x^2 = 9261,\\
&13\,104x - x^3 = 155\,520,
\end{aligned}$$

[4]Viète 1646, 178–179; 1983, 327, mentions this problem but does not give Viète's solution.

$$57x^2 - x^3 = 24\,300,$$
$$27\,755x - x^4 = 217\,944,$$
$$65x^3 - x^4 = 1\,481\,544.$$

The reason for giving the above list at length is to point out that almost all the equations are 'three-term equations'. For Viète there were several advantages to working with these relatively simple forms, the most obvious being that they require fewer lines of calculation than those with multiple affections. A more significant fact is that positively and negatively affected three-term equations have just one positive root.

'Avulsed' three-term equations, however, can yield two positive roots, and this is where Viète's treatment becomes more interesting, because to know where to start the approximation he needed to have some idea of the relative disposition of the roots. As we saw in Chapter 1, Viète had been able to use *syncrisis* to discover useful relationships between the roots and coefficients of three-term equations, and these relationships now provided him with bounds, or limits, for the two roots. This was to be so important later that it is worth pursuing a couple of examples in detail. Following Viète, we will here denote the two roots by F and G with the assumption that $F < G$.

When introducing the equation $370x - x^2 = 9261$ (in his notation $370N - 1Q$ *aequari* 9261) Viète stated that the equation has two (positive) roots and offered three conditions that must govern them: (i) one of the roots is greater than $\frac{370}{2}$, the other is less; (ii) one root is less than $\sqrt{9261}$, the other is greater; (iii) the quantity $\frac{2\times 9261}{370}$ is greater than the smaller root but less than the larger.[5] The first and second conditions are easy to explain. Viète had found by syncrisis that in equations of this type the coefficient of the linear term is $F + G$, and the 'homogene of comparison' (the term free of the unknown) is FG. In this case, therefore, he had $F + G = 370$ and $FG = 9261$, from which (i) and (ii) follow. Condition (iii), however, is not obvious, and Viète gave no explanation for it (we will return to it later). By his method of successive approximation Viète found that the smaller root is 27. It is then easy to work out that the second must be 343 (either from $370 - 27$ or $\frac{9261}{27}$), and Viète also found it by a direct application of his method.

The next example, $13\,104x - x^3 = 155\,520$ (in Viète's notation, $13{,}104N - 1C$ *aequari* 155,520), also has two positive roots. This time Viète stated the conditions on them as: (i) the square of the smaller root is less than $\frac{13\,104}{3}$, while the square of the other is greater; (ii) the quantity $\frac{3\times 155\,520}{2\times 13\,104}$ is greater than the smaller root but less than the larger.[6] Neither condition was explained, but the first can again be deduced

---

[5] *Itaque ea quae proponitur aequalitas de duobus lateribus potest explicari, quorum unum majus est semisse coefficiente, alterum minus. Immo vero unum est minus radice quadrati 9 261, alterum majus. Ac proinde cum adplicabitur duplum planum 9 261 ad 370, orietur latitudo major radice minore, minor autem radice majore.* [Thus the proposed equality may be satisfied by two roots, of which one is greater than half the coefficient, the other less. But at the same time one is less than the square root of 9261, the other greater. And further, if one divides twice 9261 by 370, there arises a quantity greater than the smaller root but less than the larger root.] Viète 1646, 211; 1983, 354.

[6] *Itaque ea quae proponitur aequalitas de duobus lateribus potest explicari, quorum unius quadratum minus est triente 13,104, alterum majus. Ac proinde cum adplicabiitur triplum solidi 155,520 ad duplum*

from Viète's earlier use of *syncrisis*. For equations of this type he had found that the coefficient of the linear term is $F^2 + FG + G^2$, and the homogene of comparison is $FG^2 + F^2G$. He therefore had $F^2 + FG + G^2 = 13\,104$ and $FG^2 + F^2G = 155\,520$. The first of these equations leads immediately to condition (i), but condition (ii) remains unexplained (again, we will return to it later). The secondary equations between F and G can be used to find either root if the other is known, and Viète did exactly that. By his method of successive approximation he determined that the smaller root of the original equation is 12. The equation $12G^2 + 144G = 155\,520$ then gave him $G = 108$. He also used his approximation method to check this directly.

Viète treated all his examples of avulsed three-term equations in similar fashion. In each case he gave instructions for calculating bounds for the roots. He also gave rules for calculating the second root from the first, and confirmed the correctness of the rules by extracting the second root directly. Nowhere, however, did he give any derivations or explanations.

With *De resolutione*, even more than with *De aeqationum*, one is left feeling that Viète had done far more work behind the scenes than he was prepared to explain. If his purpose was simply to offer a generally applicable method of solving equations, then he succeeded: his method became known as the 'general way' (*via generalis*) for solving equations, and was not superseded until late in the seventeenth century (see Chapter 9). Unexplained rules, however, appear repeatedly in the later part of the text like irritating pieces of grit, arousing both frustration and curiosity. One of Viète's earliest readers, Thomas Harriot, took it upon himself to explore and explain the rules, and in doing so was led to discoveries that permanently changed the way mathematicians thought about equations.

Harriot's unpublished treatise on equations, *c.* 1605

Nothing at all is known about Thomas Harriot's early life or background. He entered the University of Oxford in December 1577 when he was recorded as being 17 years of age, so unless he was born in the final days of December the year of his birth was 1560. A later remark by Anthony Wood suggests that he already lived in or near Oxford, but no firm trace of the family has been found. It was almost certainly at Oxford that Harriot became interested in global exploration and navigation, perhaps through the lectures of Richard Hakluyt. In 1585 Harriot joined an expedition financed by Walter Ralegh to the coast of what is now North Carolina, having already learned Algonquin from two native Americans brought back to England by an earlier expedition. His *Briefe and true report of the new found land of Virginia* (1588), written on his return, remains one of the key texts on the early European exploration of north America. During the 1590s Harriot came under the patronage of Henry Percy, ninth earl of Northumberland,

---

plani 13,104, orietur longitudo major radice minore, & minor radice majore. [Thus the proposed equality may be satisfied by two roots, of which the square of one is less than one third of 13 104, the other greater. And further when three times 155 520 is divided by twice 13 104, there arises a quantity greater than the smaller root but less than the larger root.] Viète 1646, 214; 1983, 357, mentions this problem but does not give Viète's discussion and solution.

and remained so for the rest of his life, though the Earl was imprisoned in the Tower of London from 1605 to 1621 on suspicion of association with the instigators of the gunpowder plot. Harriot lived at the Earl's London home, Syon House, and from the late 1590s onwards devoted himself to physical and alchemical experiments and to mathematics.

At the time, England was still relatively isolated from the mathematical innovations of continental Europe. The techniques of algebra were little known except through an elementary treatment in Robert Recorde's *Whetstone of witte* of 1557, and the small amount of algebra considered 'requisite for the profession of a soldiour' in Thomas Digges' *Stratiotocos* of 1579. The treatises of Viète were rare even in France; in England there can have been little knowledge of their existence, nor more than a handful of readers capable of understanding them. Harriot, however, was one such reader, and by chance was also fortunate enough to acquire most of Viète's published work. He did so through his friend Nathaniel Torporley, who in the course of his travels in the Netherlands and France met Viète in Paris and, according to later oral report, became his amanuensis.[7] A letter from Torporley to Harriot suggests that Torporley's first meeting with Viète took place in or soon after 1600.[8] It seems, therefore, that Harriot became familiar with Viète's work in the opening years of the seventeenth century and was therefore one of the first readers to subject Viète's work to careful scrutiny.[9]

In particular, Harriot read *De resolutione* in meticulous detail, re-working all of Viète's problems for himself, and adding a few more of his own.[10] His notes on Viète's 'positively affected' powers fill twelve manuscript pages. Those on Viète's 'negatively affected' powers fill a further twelve pages, in which the letter 'b' has been added to the pagination. The 'avulsed' powers are on eighteen pages marked with the letter 'c'.

The most obvious differences between Harriot's re-writing and Viète's original are changes in notation. Where Viète had used capital letters, Harriot used lower case; where Viète had written *A in B*, Harriot wrote *ab*; where Viète wrote *A-quadratus* or *A-cubus*, Harriot wrote *aa* or *aaa*; and where Viète wrote *aequari* ('is to be equalled by') Harriot used a version of the equals sign introduced by Recorde, but with two short verticals between the horizontals (to distinguish it from the sign $=$ that Viète sometimes used for subtraction). On the other hand there were similarities: Harriot, like Viète, used vowels for unknown quantities, and consonants for those given or known. He also retained Viète's concern for homogeneity, so that Viète's *Z-solido* might be

[7] 'Mr Hooke affirmes to me, that Mr Torporley was Amanuensis to Vieta: but from where he had that information he has now forgot: but he had good and credible authority for it: and bids me tell you [Anthony Wood] that it was certainly so.' Aubrey 1898, I, 263. Possibly Hooke's informant was John Pell, who around 1640 was closely acquainted with Thomas Aylesbury and Walter Warner, two of Harriot's and Torporley's former colleagues.

[8] Torporley to Harriot, 16 September [1600–1603], in BL Add MS 6788, f. 117. Torporley called Viète 'that French Apollon', which would seem to be a reference to Viète's *Apollonius gallus*, published in 1600.

[9] For a detailed analysis of Harriot's work on the treatises of Viète, see Stedall 2008.

[10] Harriot's (unlettered) pages 1 to 12 on positively affected powers are in BL Add MS 6782, ff. 388–399. Pages b.1 to b.12, on negatively affected powers, are in Petworth HMC 241.1, ff. 1–9, 11–13. Pages c.1 to c.18, on avulsed powers, are in BL Add MS 6782, ff. 400–417. The pages have been transcribed in Harriot 2003, 45–123.

written by Harriot as xxz, for example, to indicate a three-dimensional quantity that was not necessarily a cube. To the modern reader, the lack of any shorthand for repeated powers can make Harriot's expressions look rather lengthy, but on the other hand they are unambiguous and there is no difficulty in reading them. From this point on there will rarely be any need to modernize or explain notation.

We will examine first Harriot's treatment of positively affected equations, in particular his work on cubics of the form $aaa + dda = xxz$. Recall that for Harriot, a represented an unknown quantity, while dd and xxz were supposed known or given, so this is an equation involving a cube, a linear term, and a number.[11] In this context the notation dd does not mean that the coefficient of a is a square, only that it is to be regarded as a two-dimensional quantity, just as the repeated xs are also used as arbitrary dimension holders.

As an example, Harriot took up an equation already mentioned above, which Viète had written as $1C + 95{,}400N$ *aequari* 1,819,459, but which Harriot wrote as $aaa + 95{,}400a = 1{,}819{,}459$. Harriot's working is a mixture of theory and practice: first he tested that $a = 19$ does indeed satisfy the equation, but at the same time he wanted to explain why Viète's method worked. To do this he supposed that $a = b + c$, where b is a first approximation, and $b + c$ a refinement of it. Replacing a by $b + c$ in $aaa + dda = xxz$ gave him on the left hand side:

$$\begin{array}{r|l} bbb & +3bbc + 3bcc + ccc \\ +ddb & +ddc \end{array} \tag{3}$$

Harriot called this the 'canonical form' (*species canonica*) for this type of equation. The terms to the left of the vertical line, $bbb + ddb$, are those from which one should seek the first approximation. In other words, we should look for b (to the nearest ten below) such that $bbb + 95{,}400b = 1{,}819{,}459$. Clearly b must lie between 10 and 20, so we may take $b = 10$. Since $10^3 + 95{,}400 \times 10 = 955{,}000$, the remaining terms, those to the right of the vertical line, must satisfy

$$ddc + 3bbc + 3bcc + ccc = 864,459$$

or

$$95,400c + 300c + 30cc + ccc = 864,459. \tag{4}$$

Dividing 864,459 by $95,400 + 300 + 30$, as Viète had done, suggests that the next digit should be 9, and it is easily checked that this is an exact solution, so that the solution to the original equation is $a = 19$.

It is easy to see that equation (4) is the same as equation (2) earlier, and that Harriot's procedure was the same as Viète's. Their presentations of it, however, were very different. Where Viète simply gave a set of instructions couched in geometric language, Harriot introduced the a, b, c notation used above and set out his working as follows.[12] (The dots are an aid to correct alignment at each stage.)

---

[11] Harriot's work on equations of this type is reproduced in Harriot 2003, 51–52.

[12] Harriot 2003, 53.

$$
\begin{array}{lccccccccl}
 & & & & b & & c & & \\
 & 0 & & & 1 & & 9 & & \\
 & \dot{1} & 8 & 1 & \dot{9} & 4 & 5 & \dot{9} & \\
dd & & 9 & 5 & 4 & \dot{0} & \dot{0} & \cdot & \\
\hline
ddb & & 9 & 5 & \dot{4} & 0 & 0 & & \\
bbb & & & & 1 & & & & \\
\hline
ddb+bbb & & 9 & 5 & \dot{5} & 0 & 0 & & \\
\hline
ddc+3bbc+3bcc+ccc & 8 & 6 & \dot{4} & 4 & 5 & \dot{9} & & c=9 \\
\hline
dd & & 9 & 5 & 4 & 0 & 0 & & \\
3bb & & & & 3 & & & & \\
3b & & & & & 3 & & & \\
\hline
dd+3bb+3b & & 9 & \dot{5} & 7 & 3 & \dot{0} & & \\
\hline
ddc & 8 & 5 & \dot{8} & 6 & 0 & \dot{0} & & \\
3bbc & & & 2 & 7 & & & & \\
3bcc & & & 2 & 4 & 3 & & & \\
ccc & & & & 7 & 2 & 9 & & \\
\hline
ddc+3bbc+3bcc+ccc & 8 & 6 & \dot{4} & 4 & 5 & \dot{9} & & \\
\hline
 & 0 & 0 & 0 & 0 & 0 & 0 & &
\end{array}
$$

Thus where Viète annotated a line of working as, for example, 'the solid formed by the square of the side of the second and three times the first, plus the solid formed by three times the square of the first and the side of the second plus the cube of the second' Harriot was able to write simply $3bcc + 3bbc + ccc$. Not only is his notation easier to read than Viète's descriptions, but it also allows the reader to see exactly how the lines relate to each other and to the canonical form in (3). There are many examples throughout Harriot's manuscripts where his notation helps to reveal the internal structure of a problem, and this is one of them.[13]

The entire method depends, of course, on being able to make a lower estimate for the first digit. This is relatively straightforward for positively or negatively affected three-term equations, but less so for avulsed three-term equations, which have two positive roots: for these equations one must know the bounds or limits between which the roots must lie. As we have seen, Viète gave rules for finding such limits, but without explanation. Harriot was able to use symbolic manipulation not only to confirm the rules but to show how they arose. We will describe his argument as applied to the equation

[13]See also Stedall 2007.

$9261 = 370a - aa$, discussed earlier, for which Viète gave only unexplained rules.[14] Harriot described this type of equation under the general heading $xz = da - aa$. Denoting the two positive roots by b and c, he next wrote the equation in a more specific way, as $bc = ba + ca - aa$. This he called the 'canonical form for unequal roots' (*species canonica ad radices inaequales*). He had already used the description 'canonical form' (*species canonica*) in a different context earlier (see (3) above); here, however, it is clear that he was offering a general form for a quadratic equation with distinct positive roots. This was a crucial step, and we will later examine Harriot's derivation of it in greater detail.

Now suppose that b is the smaller of the roots, c the larger. We therefore have

$$2b < b + c < 2c$$

and so

$$b < \frac{b+c}{2} < c,$$

which gives Viète's condition (i):

$$b < \frac{d}{2} < c.$$

Similarly, Harriot was able to argue that

$$bb < bc < cc$$

and so

$$b < \sqrt{bc} < c,$$

which gives Viète's condition (ii):

$$b < \sqrt{xz} < c.$$

Finally, in a similar but slightly more sophisticated argument, he combined sums and products to give[15]

$$bb + bc < 2bc < bc + cc$$

and so

$$bd < 2xz < cd,$$

which gives Viète's condition (iii):

$$b < \frac{2xz}{d} < c.$$

Harriot produced one further inequality by arguing that $da > aa$ (since xz is positive) and so $d > a$.

Applying these inequalities to the equation $9261 = 370a - aa$ he therefore had

[14] See Harriot 2003, 87–91.

[15] Harriot actually gave this argument in reverse order, as an 'analysis' rather than a 'synthesis'; nevertheless, it is clear that his insight was correct.

(i) $0 < b < 185 < c < 370$,

(ii) $0 < b < 96 < c < 370$,

(iii) $0 < b < 50\frac{2}{37} < c < 370$.

Of these, (iii) gives the tightest limits for b and (i) for c. In fact, as we saw above, the solutions are $b = 27$ and $c = 343$.

A further example shows how the arguments can be extended to cubics, in particular those of the form Harriot described under the heading $xxz = dda - aaa$. This time he demonstrated his argument on the equation $155{,}520 = 13{,}104a - aaa$, which was also discussed above in relation to Viète.[16] This time Harriot gave the canonical form as

$$bbc + bcc = bba + bca + cca - aaa,$$

where, as before, b and c are the two positive roots. Again, we will return later to Harriot's derivation of this form. For now we will simply observe how he used it. From this canonical form he could see that $bb + bc + cc = dd$. As before, he supposed that b is the smaller root, c the larger, so that

$$3bb < bb + bc + cc < 3cc$$

and therefore he had

$$bb < \frac{dd}{3} < cc,$$

which gives Viète's condition (i):

$$b < \sqrt{\frac{dd}{3}} < c.$$

Similarly, he could argue that

$$2bbb < bbc + bcc < 2ccc$$

and so, since he knew that $bbc + bcc = xxz$, he had

$$bbb < \frac{xxz}{2} < ccc,$$

which leads to a condition not given by Viète:

$$b < \sqrt[3]{\frac{xxz}{2}} < c.$$

Finally, again starting from the inequality $2bbb < bbc + bcc < 2ccc$, and adding $2bbc + 2bcc$ to each part, gave him

$$2bbb + 2bbc + 2bcc < 3bbc + 3bcc < 2bbc + 2bcc + 2ccc$$

[16] See Harriot 2003, 92–98.

and so

$$b < \frac{3bbc + 3bcc}{2bb + 2bc + 2cc} < c,$$

which gives Viète's condition (ii):

$$b < \frac{3xxz}{2dd} < c.$$

As in the quadratic case, he could find an upper bound for the larger root from the condition $dda > aaa$, that is, $\sqrt{dd} > a$. Applying all these conditions in turn to the equation $155{,}520 = 13{,}104a - aaa$ gave him

(i) $0 < b < \sqrt{4{,}368} < c < \sqrt{13{,}104}$,

(ii) $0 < b < \sqrt[3]{77{,}760} < c < \sqrt{13{,}104}$,

(iii) $0 < b < \frac{466{,}560}{26{,}208} < c < \sqrt{13{,}104}$.

Harriot left the inequalities like this,[17] but to the nearest integers they can be written as

(i) $0 < b < 67, \quad 66 < c < 114$,

(ii) $0 < b < 43, \quad 42 < c < 114$,

(iii) $0 < b < 19, \quad 18 < c < 114$.

The second inequality is thus seen to be redundant. It is not difficult to show that this will always be the case, which was presumably the reason that Viète did not give it. The roots here are actually $b = 12$ and $c = 108$, and either is easily deduced from the other once one knows the composition of dd or xxz in terms of b and c.

The above analysis of inequalities for the limits of the roots sheds light not only on Viète's work but on Harriot's also, for we now see how crucial to his investigation were his 'canonical forms'. After his careful re-working of Viète's numerical examples in three manuscript sections (unlettered, 'b', and 'c'), he moved on to a fourth section, lettered 'd' and entitled 'De generatione aequationum canonicarum' ('On the generation of canonical equations').[18] It contained ideas that were to offer completely new insights into the structure of polynomial equations.

What Harriot saw was that such equations, at least where all the roots are real, can be generated by multiplying linear factors.[19] His first and simplest examples were

[17]For a similar example in his own hand see Harriot 2003, 86.

[18]See Harriot 2003, 124–164.

[19]It is possible that Harriot was influenced in this by his reading of Michael Stifel's *Arithmetica integra* (1544), a book he knew well. One of Stifel's problems is the following: *Quaero numerum mediantem inter numerũ binario maiorem, & senario minorẽ, ita ut extremi illi numeri inter se multiplicata faciant 48.* [I seek a number between a number that is larger by two, and one that is smaller by six, such that those two outer numbers multiplied together make 48.] Putting $1r$ for the number sought, Stifel multiplied together $1r + 2$ and $1r - 6$ to obtain $1z - 4r - 12$ (where $1z$ is a square), which he then set equal to 48. Stifel 1544, 277v.

the following.[20] If $a = b$ or $a = c$ then $a - b = 0$ or $a - c = 0$, and we have $(a-b)(a-c) = aa - ba - ca + bc = 0$. Throughout his work Harriot took b and c and any other unsigned letters to be positive, so this is a canonical form for a quadratic equation with two positive roots b and c. If on the other hand we have $a - b = 0$ or $a + c = 0$, then we have $(a - b)(a + c) = aa - ba + ca - bc = 0$, a different canonical form, with only one positive root, namely b. It is important to note that Harriot was here concerned with generating polynomials, not with decomposing them. At the beginning of 'De generatione' he listed the first few cases he planned to explore (the right-angled bracket indicates that the terms inside are to be multiplied):[21]

$$a-b \qquad \left.\begin{array}{c} a-b \\ a-c \end{array}\right| \qquad \left.\begin{array}{c} a-b \\ a+c \end{array}\right| \qquad \left.\begin{array}{c} a-b \\ a-c \\ a-d \end{array}\right| \qquad \left.\begin{array}{c} a-b \\ a-c \\ a+d \end{array}\right| \qquad \left.\begin{array}{c} a+b \\ a+c \\ a-d \end{array}\right|$$

For each of these in turn he derived the canonical equation. For the fifth of them, for example, the multiplication gave (as he wrote it):[22]

$$\left.\begin{array}{c} a-b \\ a-c \\ a+d \end{array}\right| = \begin{array}{llll} aaa & -baa & & \\ & -caa & +bca & \\ & +daa & -bda & \\ & & -cda & +\,bcd = 000 \end{array}$$

from which it follows that

$$bcd = \begin{array}{lll} -bca & & \\ +bda & +baa & \\ +cda & +caa & \\ & -daa & -aaa. \end{array} \tag{5}$$

Harriot checked that this equation is indeed satisfied by putting $a = b$ or $a = c$, and he also proved (by contradiction) that it is not satisfied by any other positive value of a. In other words, it is the general canonical form for a cubic with two positive roots, b and c.

The idea of constructing polynomials as products of linear factors was Harriot's outstanding contribution to the theory of equations. He also considered quadratic factors of the special kind $df + aa$, leading to $a = \sqrt{-df}$ as a possible root.[23] Further, Harriot's work showed in a visually immediate way exactly how the coefficients of an equation are composed from its roots. In equation (5) above, for instance, it is clear

[20] For the first publication of these results see Harriot 1631, 16–17; for Harriot's original manuscript version see Harriot 2003, 125–127.

[21] Harriot 2003, 125.

[22] Harriot 2003, 130–132.

[23] Harriot 2003, 158–159. Stifel too had shown how to multiply, for example, $1z + 5$ by $1z - 2$ (where $1z$ is a square); see note 19.

without any need for further explanation that the coefficient of the square term is the sum of the roots, and the coefficient of the linear term is the sum of their products in pairs. These were the kind of results that Viète had also been able to obtain, but only through his method of syncrisis, which became very cumbersome for equations of degree higher than three or four.

As we have seen, Harriot, following Viète, was particularly interested in avulsed cubic equations lacking a square term. It is clear from inspection of (5) that this arises only if $d = b + c$. If this value of d is substituted into the remaining coefficients, equation (5) reduces to

$$\begin{array}{rcll} bbc & = & bba & \\ +\,bcc & & +\,bca & \\ & & +\,cca & -\,aaa. \end{array} \tag{6}$$

This was precisely the equation Harriot had quoted as the canonical form for the particular case $155{,}520 = 13{,}104a - aaa$, and from which he had derived the various inequalities for the limits of the roots. Indeed immediately after equation (6) in 'De generatione' he referred back to that particular example.[24] From this and other cross-references it is clear that his theoretical work in 'De generatione' is very closely related to his numerical calculations earlier.

Harriot went on to investigate equations without a linear term or a square term that might arise from other cases of cubic.[25] He performed similar calculations for several fourth degree equations too. Just as for cubics, he noted the special relationships between the roots that would cause one of the terms to disappear, and calculated the remaining coefficients in terms of just three of the four roots. His most remarkable examples are those where he explored the conditions for two terms to vanish simultaneously. For example, for the fourth degree equation with roots b, c, d, and $-f$ (for by now Harriot was beginning to accept negative roots into his calculations), the necessary condition for the disappearance of the linear term is $b + c + d = f$ and of the square term $bc + bd + cd = bf + cf + df$. When both of these hold, the original equation reduces to an equation with a linear term and fourth power only, with both coefficients expressible in terms of b and c alone; in other words, an avulsed three-term equation of the kind Viète and Harriot were particularly concerned with.[26] The algebraic manipulations in this case reveal that d and f must in fact be complex, a discovery that Harriot handled without error and without comment.

One point is important to note here because it caused some confusion in the seventeenth century and has continued to do so since. Harriot was not making transformations of the kind taught by Cardano which would *cause* one (or more) of the coefficients to disappear. Rather, he was investigating special relationships between the roots, which give rise to equations in which one (or more) of the terms *does* disappear. To put it another way, he was not controlling or manipulating equations, but examining their internal structure in a series of special cases.

[24]Harriot 2003, 132.
[25]Harriot 2003, 133–139.
[26]Harriot 2003, 144.

The rest of Harriot's treatment of equations need be described here only briefly. Following 'De generatione' he wrote two further sections, lettered 'e' and 'f', thorough treatments of cubics and quartics, respectively, in which he listed and handled all possible cases, and provided numerous worked examples.[27] It is clear that he expected cubics to have three roots and quartics four, though he did not usually list repeated roots separately.[28] In the later parts of this work he routinely handled negative roots and occasionally took into account complex roots as well.[29]

Taken as a whole, the six sections of Harriot's treatise offer a systematic and detailed study of equations, full of remarkable insights and much more clearly written than anything that had gone before. He was enormously indebted to Viète in this as in several other areas of mathematics, but in comparing his work with Viète's two particularly notable achievements stand out. The first was his invention of lucid notation. Reading Harriot after his sixteenth-century predecessors one has, for the first time, the sense of looking at 'modern' mathematics. Harriot's notation was not just an improved way of writing mathematics, however: it was also an investigative tool that led him to new and significant discoveries.[30]

Harriot's other achievement was to begin a revolution in the way equations were conceived and understood. Cardano, Bombelli, Stevin, and Viète, had all regarded polynomial equations as relationships between proportional quantities. This had produced some useful insights into the structure of equations but it did not help very much in the more practical matter of solving them. Harriot's treatment of polynomials as products of factors opened up a range of new insights. His generation of canonical forms showed immediately, for instance, that an equation could be expected to have as many roots as its degree, and also made clear how the coefficients were constructed from the roots. He extended his work only occasionally and partially to complex roots: he experimented, for instance, with quadratic factors of the form $aa \pm df$, but for some reason never the form $aa \pm ba \pm df$. Nevertheless, his treatise was, for its time, a highly original and innovative piece of work.

Up to two centuries after his death Harriot's achievement was well recognized. Charles Hutton, for instance, in the entry for 'Algebra' in his *Mathematical and philosophical dictionary* of 1795–96, wrote:[31]

> [Harriot] shewed the universal generation of all the compound or affected equations, by the continual multiplication of so many simple ones, or binomial roots; thereby plainly exhibiting to the eye the whole circumstances of the nature, mystery and number of the roots of equations; with the composition and relations of the coefficients of the terms; and from which many of the most important properties have since been deduced.

---

[27] Harriot 2003, 174–286.

[28] See, however, Harriot 2003, 233.

[29] For one of the best examples of his use of complex roots see Harriot 2003, 237.

[30] See Stedall 2007.

[31] Hutton 1795–96, I, 96. The same paragraph is also in Hutton 1812, II, 286.

16 SECTIO SECVNDA.

Canonicarum quadratici ordinis deriuatio.

PROPOSITIO PRIMA.

Aequatio canonica $\begin{matrix} aa - ba \\ + ca \end{matrix} = + bc$ ab originali $\begin{matrix} a - b \\ a + c \end{matrix} = \begin{matrix} aa - ba \\ + ca - bc \end{matrix}$ posito b. ipsi a. æquali deducitur.

Nam si ponatur $a = b$. erit $a - b = o$.
Posito igitur $a = b$. erit $\begin{matrix} a - b \\ a + c \end{matrix} = o$.
Est autem ex genesi $\begin{matrix} a - b \\ a + c \end{matrix} = \begin{matrix} aa - ba \\ + ca - bc \end{matrix}$. quæ est æquatio originalis hic designata.
Ergo... $\begin{matrix} aa - ba \\ + ca - bc \end{matrix} = o$.
Ergo... $\begin{matrix} aa - ba \\ + ca \end{matrix} = + bc$. quæ est æquatio proposita.
Aequatio igitur canonica proposita ab originali designata, posito b. ipsi a. æquali deducitur. Vt est enunciatum.

PROPOSITIO 2.

Aequatio canonica $\begin{matrix} aa - ba \\ - ca \end{matrix} = - bc$. ab originali $\begin{matrix} a - b \\ a - c \end{matrix} = \begin{matrix} aa - ba \\ - ca + bc \end{matrix}$. posito b. vel c. ipsi a. æquali deducitur.

Nam si ponatur $a = b$. erit $a - b = o$.
Vel . . . $a = c$. erit $a - c = o$.
Posito igitur $a = b$. vel c. erit $\begin{matrix} a - b \\ a - c \end{matrix} = o$.
Est autem ex genesi $\begin{matrix} a - b \\ a - c \end{matrix} = \begin{matrix} aa - ba \\ - ca + bc \end{matrix}$. quæ est æquatio originalis hic designata.
Ergo . . . . $\begin{matrix} aa - ba \\ - ca + bc \end{matrix} = o$.

Ergo

A new understanding of equations (2): polynomials as products of factors, from Harriot's *Praxis* (1631).

Unfortunately, from about 1800 onwards, Harriot's work fell into oblivion. He had not published his discoveries himself in his lifetime and it was his friend and colleague Walter Warner who eventually edited some of them posthumously in the *Artis analyticae praxis* (1631). Warner never had Harriot's deep understanding of the subject, however, and simply omitted whatever he found obscure or difficult. Thus, in the *Praxis*, there are no negative or complex roots, Harriot's use of coefficients to calculate upper and lower bounds is omitted, and his investigations of three-term quartics are abandoned in a fog of incomprehension.[32]

Not only is the careful structure of Harriot's work on equations all but lost in the *Praxis* but so, unfortunately, is the motivation for it. From his unpublished manuscripts it is clear that Harriot first worked on numerical solution, which in turn led him to investigate the information that could be deduced from the coefficients, and so to make general observations about the coefficients in relation to the roots. In the *Praxis*, however, the reader meets only repetitive manipulations and lists of canonical equations, all of them divorced from the primary problem of equation-solving. A few worked examples are thrown in at the end of the book, but bear little relation to anything that has gone before. The *Praxis* therefore did less than justice to the skill and subtlety of Harriot's original work. In seventeenth-century English mathematical circles the *Praxis* was always mentioned with respect, but most readers can only have been somewhat bemused by its contents. Until the end of the twentieth century, however, it was the book upon which Harriot's reputation rested.

Girard's *Invention nouvelle en l'algebre*, 1629

Two years before the *Praxis* was posthumously published, several useful properties of the coefficients of polynomial equations were explored in a treatise of a quite different kind, Albert Girard's *Invention nouvelle en l'algebre*. Girard appears to have come from St Mihiel, close to the modern French–Belgian border, but to have spent most of his life in the Netherlands. Like Stevin some thirty years earlier, he was an engineer in the Dutch army (Stevin had served under Maurice of Nassau, Girard served under Maurice's younger brother Frederik Hendrik). Indeed Girard was thoroughly familiar with the mathematical writings of Stevin, some of which he edited as *Les oeuvres mathématiques de Simon Stevin*.[33]

Girard's *Invention nouvelle* was not a theoretical text. Rather, after a good deal of preliminary discussion of arithmetic, Girard offered practical instructions, with worked examples, for solving quadratic and cubic equations. One has a strong sense in reading the book that Girard made new discoveries even as he was writing, and that he then simply incorporated these into the next part of his text. Thus after giving the rules for quadratics and cubics, Girard turned to 'a new way of solving the said equations' (*une nouvelle maniere pour resoudre les susdites equations*), possibly the *nouvelle invention*

[32]Harriot 1631, 46; Harriot 2007, 63.

[33]*Les oeuvres* includes Stevin's *L'arithmetique … aussi l'algebre*, and Stevin's translation of Books I–IV of the *Arithmetic* of Diophantus, to which Girard added translations of Books V and VI. It also contains several treatises on the mathematical sciences. It was published in 1634, two years after Girard's death.

of his title. He began with the example 1② *esgale à* 6① + 40 (in modern notation $x^2 = 6x + 40$). Division by 1① gives 1① *esgale à* $6 + \dfrac{40}{1①}$. If the root 1① is an integer, then it must be the case that it divides 40 exactly, and it does not take long to test the divisors of 40 to find that the required value of 1① is 10. Similarly, the equation 1③ *esgale à* 7① – 6 can be solved by searching for divisors of 6 with the property that 1② *esgale à* $7 - \dfrac{6}{1①}$. In this case Girard noted that each of 1, 2, and -3 is a possible value of 1①. This and similar findings led him to the following conjecture: that equations have as many roots as the degree of the highest power indicates, unless they are incomplete, that is, where one or more of the terms is 'missing'.[34] Almost immediately, in trying to explain why incomplete equations are an exception, Girard saw that a 'missing' term is simply a term with a zero coefficient. This seems to have caused him to revise his theorem, because by the end of the passage, two pages later, he is convinced that *every* equation has as many roots as its degree, including repetitions and complex roots if necessary.[35] Thus he lists the roots of 1④ *esgale à* 4① – 3 as $1, 1, -1 + \sqrt{-2}, -1 - \sqrt{-2}$, whose sum is zero and whose product is 3. The use of such 'impossible' solutions, says Girard, is that they make the rule for the number of roots quite general and ensure that no root is missed.

Girard also saw that much useful information could be deduced from the coefficients. For a given set of roots he defined their 'first faction' to be their sum; their 'second faction' to be the sum of their products two at a time; their 'third faction' to be the sum of their products three at a time; and so on.[36] His preferred way of writing equations was with even powers on the left hand side (with a leading coefficient of 1) and odd powers on the right. Under these conditions he was able to claim that the 'factions' are simply the coefficients of the terms, from the second highest power downwards.[37] We do not know Girard's sources, but it is possible that this insight came from reading Viète's *Tractatus duo*. He himself offered no proof or justification of his assertion but was able to put it to good use. Given the equation 1③ *esgale à* 300① + 432, for example, one can discover by testing divisors of 432 that one of the roots is 18. This means that the sum of the other two roots is -18 and their product is 24, that is, they must satisfy 1② *esgale à* $-18$① – 24. This is just a quadratic equation, easily solved to give the second and third roots, $-9 + \sqrt{57}$ and $-9 - \sqrt{57}$.

Finally, denoting the coefficients of the terms after the first by A, B, C, and so on, Girard claimed (without proof) that for any equation

[34] *Toutes les equations d'algebre reçoivent autant de solutions, que la denomination de la plus haute quantitié le demonstre, exceptè les incomplettes:* [All equations in algebra have as many solutions as the degree of highest term indicates, except for those that are incomplete.] Girard 1629, Theorem II, sigs [E4]–[E4]v.

[35] *Donc il se faut resouvenir d'observer tousjours cela.* [Therefore one must remember to note this in every case.] Girard 1629, sig F.

[36] Girard 1629, Definition XI, sigs [E3]v–[E4].

[37] Girard 1629, Theorem II, sig [E4]v.

the sum of the solutions is	A,
the sum of their squares is	$A^2 - 2B$,
the sum of their cubes is	$A^3 - 3AB + 3C$,
the sum of their fourth powers is	$A^4 - 4A^2B + 4AC + 2B^2 - 4D$.

It is clear that Girard had a great deal of insight into the composition of the coefficients of polynomial equations, but he offered his assertions without any explanation. He never wrote equations with all the terms set equal to zero, for instance, and if he had any idea that polynomials might be factorized he gave no hint of it. In other words, he came up with many of the same insights as Harriot as to the number of the roots and the nature of the coefficients but, as far as we can see, without any theoretical underpinning. The theory was to appear in the *Praxis* just two years after the publication of *Invention nouvelle*. Girard, however, died in December 1632 at the age of 37, and was probably never aware of Harriot's work.

Descartes' *La géométrie*, 1637

Far more influential than either Girard's *Invention nouvelle* or Harriot's *Praxis* was a book that appeared just a few years after them, *La géométrie* of René Descartes. Published in 1637 as an appendix to Descartes' *Discours de la méthode*, it proved to be one of the seminal texts of seventeenth-century mathematics, its fundamental themes being the analysis of geometric problems by means of algebra, and the geometric construction of the solutions.[38] Descartes' treatment of equations occupied only a few pages,[39] but like everything else in *La géométrie* gave rise to a great deal of further discussion.

Descartes treated equations from the start as a collection of terms equal to zero.[40] As Harriot had done, and indeed in much the same language, he showed how a polynomial can be constructed from its roots, but did so only by means of a single numerical example. Thus, he claimed, if $x-2=0$ or $x-3=0$ or $x-4=0$ or $x+5=0$, then the appropriate equation will be $(x-2)(x-3)(x-4)(x+5)=0$, or (in Descartes' notation) $x^4 - 4x^3 - 19xx + 106x - 120 = 0$. The question of whether Descartes was influenced in this by Harriot remains unresolved and probably unresolvable: for further discussion on the matter see below.

Almost immediately Descartes then stated a rule that was to lead to great deal of investigation later:[41] that the number of positive roots (*racines vrayes*) may be as many as the changes of sign from $+$ to $-$ or from $-$ to $+$; and the number of negative roots (*racines fausses*) may be as many as successions of the same sign, whether from $+$ to $+$ or $-$ to $-$. Although Descartes expressed the rule in terms of the number of roots

[38] See Bos 2001.
[39] Descartes 1637, 372–387.
[40] Descartes 1637, 372–374.
[41] *A sçavoir il y en peut auoir autant de vrayes, que les signes + & − s'y trouuent de fois estre changés; & autant de fausses qu'il s'y trouue de fois deux signes + ou deux signes − qui s'entresuiuent.* [That is, one may have as many true roots as the number of times the signs + and − are found to change; and as many false roots as the number of times two + signs or two − signs follow each other.] Descartes 1637, 373.

that *may* be found, his example suggested something more precise. In the equation $[+]x^4 - 4x^3 - 19xx + 106x - 120 = 0$, where the sign pattern is $+ - - + -$, we can expect, according to the rule just stated, up to three positive roots and one negative. Descartes, however, claimed more than that:[42] 'one knows that there *are* three true roots and one false' (my italics). In this case, because all the roots are real, the rule gives the actual rather than potential number of positive and negative roots, but Descartes did not explain why this should be so, or indeed anything else concerning it.

Descartes also gave some basic rules for transforming equations.[43] He pointed out, for instance, that changing the signs of the odd powers in an equation is equivalent to changing positive roots to negative, and vice versa. Further, to increase each root by 3, say, we should use the transformation $y = x + 3$. He claimed two particular uses of this technique: (i) to remove the second highest term and (ii) to increase the roots by a sufficiently large amount to ensure that all the roots of the new equation will be positive. He also noted that it is possible to eliminate fractions and surds by appropriate multiplication of the roots. All of this was standard technique and by now well known.

To solve cubic equations, Descartes suggested that one should search for a root by inspecting divisors of the term free of the unknown. (Girard had done the same but Descartes did not mention it; as in relation to Harriot it is impossible to know whether Descartes was influence by Girard or not.) If α, say, is such a divisor, then one should test whether $x - \alpha$ divides the polynomial.[44] This could be tried on quartics too, but here Descartes had another idea.[45] First, remove the cube term so that the equation takes the form (as Descartes wrote it): $+x^4 * .pxx.qx.r = 0$ (in modern notation $x^4 \pm px^2 \pm qx \pm r = 0$). If we suppose that the expression on the left is a product of two quadratic factors, we must have, for appropriate values of y, f, and g,

$$x^4 \pm px^2 \pm qx \pm r = (x^2 - yx + f)(x^2 + yx + g). \tag{7}$$

Multiplying out, and equating coefficients, yields the three equations

$$\begin{aligned} \pm r &= fg, \\ \pm q &= fy - gy, \\ \pm p &= f + g - y^2, \end{aligned}$$

and elimination of f and g gives rise to a cubic equation in y^2, namely,

$$y^6 \pm 2py^4 + (p^2 \pm 4r)y^2 - q^2 = 0. \tag{8}$$

Once a value of y is found from (8), f and g are easily calculated. Note that the Cardano–Ferrari method for quartics also gives rise to an intermediate cubic, but of a different and simpler kind than the cubic that arises in Descartes' method.

---

[42] *on connoist qu'il y a trois vrayes racines; & vne fausse, a cause que les deux signes* $-$, *de* $4x^3$, *&* $19xx$, *s'entresuiuent.* [one knows that there are three true roots and one false because the two $-$ signs, of $4x^3$ and $19xx$ follow each other.] Descartes 1637, 373.

[43] Descartes 1637, 374–380.

[44] Descartes 1637, 380–383.

[45] Descartes 1637, 383–387.

Descartes offered equation (8) without any explanation, and not until two pages later did he show how the value of y can be used as shown in (7) to write the original quartic as the product of two quadratic factors. It is not surprising that at least one of his early readers was very baffled: Sir Charles Cavendish wrote to John Pell in 1646 fearing that Descartes' text was 'fals printed', and begging for an explanation. Although Pell tried to help him, Cavendish was still struggling with the matter three years later, at which point Pell wrote out a full and systematic explanation.[46] John Wallis almost certainly drew on Pell's work when he too demonstrated the method in 1685 in *A treatise of algebra*. He did so because, he complained, 'How he [Descartes] came by that Rule he doth no where tell us'.[47]

Descartes claimed that he could give rules for equations of degree five, six, or higher but preferred to say only that in general one should approach such equations by trying to write them as a product of two others of lower degree;[48] if this proved impossible then one had to turn instead to solution by geometric construction.

The only completely new results in Descartes' treatment of equations were (i) his rule of signs, which set upper bounds for the number of positive or negative roots and (ii) his method for solving quartics. Otherwise, the various transformations he prescribed had all been known since Cardano, and the method of composing polynomials as products of factors had been thoroughly explored by Harriot. Thus, controversy arose almost as soon as *La géométrie* was published, and rumbled on for a long time afterwards, as to whether Descartes had taken results from Viète and Harriot without acknowledgement. Descartes denied that he had read the work of either, a denial that raises subtle questions about mathematical precedence. The work of Viète had been circulating in France for more than 40 years and it is hardly conceivable that Descartes was unaware of it. Likewise, Harriot's *Praxis* was known in Paris during the 1630s, and even oral report of its contents would have been enough for Descartes to reconstruct for himself the idea of polynomials as products of factors. On the other hand, we may have here an example of a phenomenon that is by no means unknown in mathematics, namely, the discovery of similar results within a relatively short span of time by mathematicians working quite independently.

Hutton in 1795 had recognized Harriot's priority and the importance of his contribution, but during the nineteenth and twentieth centuries Harriot's achievements, presented in tedious and unattractive style in the *Praxis*, became overshadowed by those of Descartes, whose work was by then so much better known and visible. Thus, to take but one example from a later historian,[49] John Stillwell in 2002 claimed that an important contribution made by Descartes was 'the theorem that a polynomial $p(x)$ with value 0 when $x = a$ has a factor $(x - a)$'. We may ignore the anachronisms of modern notation and the point that Descartes presented only one example, not a theorem. We cannot ignore, however, the fact that Harriot had begun a systematic

[46]See Malcolm and Stedall 2005, 473–474, 535, 294–295; Pell's treatment of the problem is in BL MS Harleian 6083, ff. 100v–101.

[47]Wallis 1685, 208–212.

[48]Descartes 1637, 389.

[49]Stillwell 2002, 97.

exploration of the factorization of polynomials shortly after 1600 and that his essential findings were published in 1631, six years before *La géométrie*. All of which goes to show how easily history can be re-written. The truth is that solution of equations was never Descartes' primary concern, but because *La géométrie* was so influential, the few comments on equations that Descartes made there came to dominate all further research.

Chapter 3
From Descartes to Newton

The work of Harriot and Descartes described in the previous chapter transformed for ever the way polynomial equations were studied. After the 1630s, the idea of equations as proportional relationships was replaced by the more fruitful concept of polynomials as products of factors, and notions of proportion disappeared rapidly and almost completely from seventeenth-century algebra texts.

Remaining advances during the seventeenth century were on a smaller scale: techniques for detecting double roots, thereby reducing the degree of an equation, from Jan Hudde; the identification of certain higher degree equations that were easily solvable, by François Dulaurens; a half worked out proposal for the removal of intermediate terms, from Walter von Tschirnhaus; rather better worked out insights from James Gregory and Leibniz, but which unfortunately remained unpublished and invisible; and a new way of visualizing polynomials as curves with respect to co-ordinate axes from Isaac Barrow and John Collins. In the short term none of these ideas led to significant developments, but all were to be important when equations became more intensively studied during the eighteenth century.

The final author discussed in this chapter is Isaac Newton, whose *Arithmetica universalis* was published in 1707. Although the book appeared in the early years of the eighteenth century, it was so firmly rooted in the algebra of the seventeenth that it properly belongs in this chapter as a last word on the theory of equations up to the end of the seventeenth century.

The extended *Geometria*, 1659–1661

Within a few years of its publication, Descartes' *La géométrie*, originally written as an appendix to his *Discours*, was translated into Latin by Frans van Schooten and republished in its own right under the title *Geometria* in 1649. Ten years later van Schooten brought out a second edition, now expanded to two volumes by the commentary and research that had already accumulated around Descartes' text, much of it from van Schooten himself or from his pupils, such as Jan Hudde and Hendrik van Heuraet. Four treatises in this second edition were particularly concerned with equations, two by Florimond de Beaune and two by Jan Hudde.

Florimond de Beaune was trained in law, which he practised for most of his life in his home town of Blois in the Loire valley. In his spare time he did mathematics and wrote explanatory 'Notae breves' ('Brief notes') to accompany the first Latin edition of Descartes' *Geometria* in 1649. He died in 1652 but his two treatises on equations, 'De natura aequationum' and 'De limitibus aequationum' were published posthumously in volume II of the second edition.[1] A brief description of their contents is included here for completeness, but neither contained anything that was by then new or remarkable. 'De natura' was an extended exploration of the construction of polynomials as products of factors. De Beaune gave rules for all cases of polynomials up to degree four. All his cubics were formed from multiplication of $x - b$ by a term of the form $xx \pm cx \pm dd$ (where dd represented a two-dimensional term, not necessarily a square).[2] For quartics, however, he restricted himself to multiplying a linear factor $x-b$ by a cubic factor of the form $x^3 \pm cx^2 \pm ddx + f^3$, thus ignoring Descartes' idea of using two quadratic factors. His method helped de Beaune to derive a few results concerning the composition of the coefficients in terms of the roots. All of this had appeared in the *Invention nouvelle* and the *Praxis* but de Beaune presented it at greater length and more explicitly. In his second treatise, 'De limitibus aequationum' he gave rules for finding limits, or bounds, for the roots for each case of quadratic, cubic, or quartic equations. All the rules were based on the assumption that the roots are real and positive: thus, for example, de Beaune argued that for the equation $x^3 - mmx + n^3 = 0$, it must be the case that $x > \frac{n^3}{mm}$ (since $x^3 > 0$). Since Viète's findings on this subject were obscure and Harriot's were unpublished, de Beaune's treatment at least offered useful and transparently explained starting points for finding limits.

More important than de Beaune's writings were two treatises by Jan Hudde, published in Volume I of the second edition of the *Geometria*, and entitled 'De reductione aequationum' and 'De maximis et minimis'.[3] Hudde had learned mathematics under Frans van Schooten while he was pursuing law studies in Leiden from about 1648 onwards, and probably continued to study with him afterwards. All his mathematical output comes from a period of ten years between 1654 and 1663; after that he became caught up in civic administration, and eventually became one of the four burgomasters of Amsterdam. 'De reductione' appears in the *Geometria* in the form of a letter to van Schooten, sent on 15 July 1657 but based on work done some years earlier. The letter was originally in Dutch but was translated into Latin by van Schooten.

Hudde's 'De reductione' was the first published commentary on Descartes' remark that the best approach to equations of fifth or sixth degree was to seek to write them as products of two equations of lower degree (see page 47). Hudde explored at length the factorization (*reductione*) of polynomials with literal coefficients, including those where the coefficients may be fractions or surds. Suppose, for instance, we seek a quadratic divisor of the form $xx + yx + aa$ for

$$x^4 - 2ax^3 + 2aaxx - 2a^3x + a^4 = 0. \tag{1}$$

[1] Descartes 1659–61, II, 49–116 and 117–152.

[2] Descartes introduced the notation x^3, x^4, ... for powers of x beyond a cube, but for some reason it remained customary in the seventeenth century, and even into the eighteenth, to write xx for what we would now write as x^2. We will retain the original usage in this chapter wherever seventeenth-century sources are quoted.

[3] Descartes 1659–61, I, 407–506 and 507–516.

By substituting $xx = -yx - aa$ into (1), Hudde arrived at[4]

$$(-y^3 - 2ayy + ccy)x - aayy - 2a^3y + aacc = 0.$$

This would be satisfied, he argued, provided both the following hold:

$$-y^3 - 2ayy + ccy = 0 \tag{2}$$

and also

$$-aayy - 2a^3y + aacc = 0. \tag{3}$$

Now (2) and (3) are both satisfied if $-yy - 2ay + cc = 0$, that is, if

$$y = -a \pm \sqrt{aa + cc}.$$

A divisor of the original equation (1) is therefore

$$xx - ax \pm \sqrt{aa + cc}x + aa.$$

In this example, it is easy to see the common divisor of (2) and (3). Hudde later offered a method for finding common divisors where they are not so obvious (we will examine that shortly). Before that, however, he gave the rule for which he was to be best remembered, for discovering repeated roots.

Suppose a polynomial equation has a repeated root α. Hudde observed that if the terms of the equation are multiplied by successive terms of an arithmetic progression, the new equation will also have the root α. Consider, for example,[5]

$$x^3 - 4xx + 5x - 2 = 0. \tag{4}$$

Multiplying the terms by 3, 2, 1, 0, respectively gives a new equation

$$3x^3 - 8xx + 5x * = 0, \tag{5}$$

where the symbol * (for Hudde as for Descartes) indicated an absent term. Searching for roots that equations (4) and (5) have in common we find that both equations are satisfied when $x = 1$. Therefore, Hudde claimed, 1 is a *double* root of (4).

Using differential calculus it is easy to prove that a double root of a polynomial equation is also a root of its first derivative, and one can see that what Hudde did in moving from (4) to (5) was essentially a process of differentiation. It is not so easy to see why the method works for *any* arithmetic progression, either decreasing or increasing. In delivering the rule Hudde gave no explanation, but offered a partial proof in his next treatise, 'De maximis and minimis'.[6] There he proved that if an equation of the form

$$(x - y)^2(x^3 + pxx + qx + r) = 0, \tag{6}$$

[4]Descartes 1659–61, I, 427–428.
[5]Descartes 1659–61, I, 434.
[6]Descartes 1659–61, I, 507–509.

which has a double root y, is multiplied term by term by an arbitrary arithmetic progression a, $a \pm b$, $a \pm 2b$, $a \pm 3b$, ..., then the new equation will also have y as a root. He did so by first considering just $(x - y)^2 = 0$. Multiplying xx, $-2xy$, yy respectively by a, $a + b$, $a + 2b$, he formed the new equation

$$axx - (a + b)2xy + (a + 2b)yy = 0. \tag{7}$$

Equation (7), like equation (6), is satisfied by putting $x = y$ since $a - 2(a + b) + (a+2b) = 0$. Hudde pointed out that the same still holds if (7) is multiplied through by x^3, pxx, qx, or r. Thus 'Hudde's rule' holds for equation (6) and therefore for any fifth (or higher) degree equation with a double root. It is not difficult to convince oneself that the same argument will hold for roots with higher multiplicity, which Hudde stated but did not prove.

Hudde's method of finding roots in common is illustrated here using one of his own examples.[7] For some reason he here abandoned the use of x for the unknown, and instead chose to work with two equations in d, namely,

$$d^3 - add + 2aab - 2abd = 0 \tag{8}$$

and

$$d^4 - bbdd + aabb - aadd = 0. \tag{9}$$

Suppose that (8) and (9) have a root in common. Then any equation formed from them by adding multiples of one to the other will have that same root. Hudde did not explain this, but his working suggests what he had in mind. First he multiplied (8) by d, which gives

$$d^4 = ad^3 - 2aabd + 2abdd.$$

Substituting for d^3 from (8) and simplifying, we have

$$d^4 = +aadd - 2a^3b + 2abdd.$$

Now substituting for d^4 from (9) and simplifying, we have

$$aabb - 2a^3b + 2abdd - bbdd = 0,$$

which is satisfied when

$$dd = aa \tag{10}$$

or (since Hudde was interested only in a positive root)

$$d = a.$$

Now Hudde could check that when $d = a$ both (8) and (9) are also satisfied.

Essentially he had used (8) and (9) to eliminate d^4 and d^3 and so to end up with a quadratic equation in d. There is a problem here, however, that he did not seem to

[7] Descartes 1659–61, 422.

notice. Equation (10) also gives $d = -a$, but this value of d satisfies only (9) and not (8). The fact that the method can throw up superfluous roots became well recognized later, but it does not appear to have worried Hudde.

Most of the remainder of Hudde's treatise was taken up with his initial purpose of finding and listing possible divisors for equations of degree 4, 5, or 6 with literal coefficients. Towards the end, he examined the question of whether in general an equation of the form

$$x^6 * qx^4 + rx^3 + sxx + tx + v = 0 \tag{11}$$

(that is, with no term in x^5) can be factorized as a product of a quartic and a quadratic as[8]

$$(x^4 - yx^3 + zxx + kx + l)(xx + yx + w) = 0. \tag{12}$$

By multiplying out, and equating coefficients between (11) and (12), Hudde was able to eliminate in turn z, k, l, and w, but was then left with an equation in y of degree 15. He also tried factorizing (11) as a product of two cubics, but that turned out to be even worse, leading to an equation in y of degree 20. Replacing (11) by an equation of degree 5 (with no term in x^4) led him to an equation in y that was only of degree 10, but nevertheless unsolvable. Only for an equation of degree 4 (with no term in x^3) did the method work: now the equation in y was of degree 6 but contained only terms in yy and so was essentially of degree 3. In this case the factors obtained were those that Descartes had also found, as Hudde immediately noted.

At the end of 'De reductione' Hudde turned finally to cubic equations.[9] He argued that the problem of solving a quartic equation can always be reduced to solving a cubic, and since the square term can always be removed from a cubic, the key procedure (for both cubics and quartics) is to solve cubics of the form $x^3 = * \pm qx \pm r$. The method he suggested was to put $x = y + z$, so that the equation $x^3 = qx + r$, for example, becomes

$$y^3 + 3yyz + 3yzz + z^3 = qy + qz + r. \tag{13}$$

Then he separated (13) into the two equations

$$3zyy + 3yzz = qy + qz \tag{14}$$

and

$$y^3 + z^3 = r. \tag{15}$$

This separation seems somewhat arbitrary. After all, why should (14) and (15) hold separately from (13)? On the other hand it is certainly true that if (14) and (15) hold then so does (13). Hudde gave no explanation on this point, but from equations (14) and (15) he obtained

$$y^3 = \frac{r}{2} \pm \sqrt{\frac{1}{4}rr - \frac{1}{27}q^3}$$

[8]Descartes 1659–61, 487–490.

[9]Descartes 1659–61, I, 499–501.

and

$$z^3 = \frac{r}{2} \mp \sqrt{\frac{1}{4}rr - \frac{1}{27}q^3}.$$

Putting $x = y + z$ then gave him Cardano's rule.[10]

Hudde's second and much shorter treatise, 'De maximis et minimis', is dated 6 February 1658, a few months after 'De reductione'. Hudde knew from Fermat's earlier treatment of the subject that a maximum or minimum is indicated by a double root.[11] Hudde began 'De maximis et minimis' with the proof described above of his rule for double roots, and applied the rule almost immediately to finding the maximum or minimum value of a polynomial.[12] Suppose, for example, a maximum or minimum is denoted by z. Then the equation that arises from setting the polynomial equal to z will have a double root. Thus, for example, to find a maximum or minimum value z of $3ax^3 - bx^3 - \frac{2bba}{3c}x + aab$, set

$$3ax^3 - bx^3 \ * \ -\frac{2bba}{3c}x + aab - z = 0 \tag{16}$$

and multiply each term by the exponent of x, that is, by 3, 3, 2, 1, 0, 0 to give

$$9ax^3 - 3bx^3 \ * \ -\frac{2bba}{3c}x = 0,$$

or

$$9axx - 3bxx - \frac{2bba}{3c} = 0. \tag{17}$$

A solution to (17) substituted back into (16) will give the required value of z. Hudde was aware from his previous work that any arithmetic progression can be used to derive an equation like (17) from (16), but he chose to use the progression corresponding to the degree of each term so that terms in which x does not appear conveniently vanish. Differential calculus, of course, gives (17) as the first derivative of (16), but Hudde's work was entirely algebraic, and involved no infinitesimal or limiting processes.

Because the *Geometria* was so widely read, Hudde's results became well known, and were highly regarded by several later writers.

The removal of terms, 1667–1683

In 1667 the Parisian textbook writer François Dulaurens made a throwaway remark whose significance he himself can barely have grasped. It was to be the first impulse, however, behind a wave of activity that continued through the next decade, generating new methods and techniques and drawing in mathematicians of the stature of James Gregory and Gottfried Wilhelm Leibniz. In the end it all died away, mostly because the technical difficulties proved insuperable, leaving behind incomplete ideas languishing

[10] Hudde's y, z are equivalent to $\sqrt[3]{u}$, $\sqrt[3]{-v}$ on page 7.

[11] See Chapter 1, note 49.

[12] Descartes 1659–61, 509–515.

unread in private correspondence. Few traces of this upsurge of interest in equation-solving were visible to later generations, who had to rediscover many of the ideas for themselves, yet it produced some wonderfully rich mathematics as well as some small human dramas.

I have been unable to discover anything of the life of François Dulaurens except that during the late 1660s he lived in Paris, where he was acquainted with the arithmetician Frenicle de Bessy and the scholar and bibliophile Henri Justel.[13] The dedicatory letter of his *Specimina mathematica*, published in Paris in 1667, suggests connections in Belgium, but otherwise the details of his life are obscure.[14] In 1676 Henry Oldenburg wrote of Dulaurens in the past tense and referred to surviving papers, which suggests that by then he had died.[15]

The *Specimina* is for the most part an elementary textbook, written in two parts. Book I deals with rules for proportions, and geometry. Book II treats equations: the first four of its five chapters offer introductory material and the standard rules for quadratics, cubics, and quartics, respectively. These chapters are written in a mixture of Descartes' notation and Harriot's, but are underpinned by Viète's concept of equations as proportional relationships, and so by 1667 must already have seemed rather old-fashioned. The most significant chapter of the *Specimina*, however, is the fifth and last chapter of Book II, where Dulaurens turned his attention to equations of degree higher than four. Here he identified a special class of equations that he could solve by inspection of the coefficients, without recourse to numerical methods, or to geometric construction, or to factorization. The equations in question are all related to angle division, and it is likely that Dulaurens had first come across them in Viète's *Ad angularium sectionum analyticen theoremata* (*Towards an analytic theory of angle division*), written in 1591 and completed and published by Alexander Anderson in 1615. Where Viète and other writers had solved such equations trigonometrically, however, Dulaurens saw how to solve them algebraically.

It had long been known that the problem of trisecting an angle θ gives rise to an equation of degree 3, of the form $c = 3x - x^3$. (The equation is easily obtained from the identity $\sin 3\theta = 3\sin\theta - 4\sin^3\theta$ by putting $\sin\theta = \frac{x}{2}$ and $\sin 3\theta = \frac{c}{2}$.) Viète had derived the equation in 1593 in his *Supplementum geometriae*,[16] and had shown how to solve such equations numerically in *De resolutione*. Girard had also shown in his *Invention nouvelle* how to solve cubics lacking a square term, using tables of sines, as had Pierre Hérigone in the final volume of his *Cursus mathematicus*.[17]

Equations for the division of angles into five or more equal parts were also well known. Viète had given equations for division into up to nine parts, with the de-

[13]See Hall and Hall 1965–86, IV, letter 859 (Dulaurens to Oldenburg) and, for example, letters 739, 860, 870, 919 (Justel to Oldenburg).

[14]The dedicatory letter is addressed to 'Domini ordines generales foederati belgii'. There is no entry for Dulaurens in the *Biographie universelle*, the 45-volume biographical dictionary edited by Louis Gabriel Michaud, Paris, 1842–65.

[15]Oldenburg to Leibniz, 26 July 1676, in Oldenburg 1965–86, XIII, 6.

[16]Viète 1646, 248–249, 256–257; 1983, 403–404, 416–417.

[17]Girard 1629, unpaginated; Hérigone 1644, 42–44.

gree of the equation corresponding in each case to the number of parts.[18] Indeed, in his *Responsum ad problema, quod* [...] *proposuit Adrianus Romanus* (*Response to a problem proposed by Adriaan van Roomen*) in 1595 he had shown how to solve such equations trigonometrically for degrees three, five, and, famously, forty-five.[19] Henry Briggs also had given equations for trisection, quinquisection, and septisection in his *Trigonometria britannica*, posthumously published in 1633.[20]

In the final chapter of Book II of the *Specimina*, Dulaurens too derived equations for dividing angles into 2, 3, 4, 5 or 7 equal parts. Here is his argument for quinquisection. Suppose that an angle (which we may call 5θ) is subtended by a chord of length g at the centre of a circle of radius r (so that $g = 2r \sin 5\theta$). According to Dulaurens, one-fifth of the angle is subtended by a chord of length a where[21]

$$a^5 - 5r^2a^3 + 5r^4a - r^4g = 0. \tag{18}$$

Next Dulaurens had the idea, which he described as *per mesolabum* (by two mean proportionals), of setting $a = m + n$. Expansion of $(m+n)^5$ shows that

$$a^5 - 5mna^3 + 5m^2n^2a - m^5 - n^5 = 0. \tag{19}$$

A general equation of the form

$$a^5 - qa^3 + sa - t = 0$$

can therefore be solved if we can find m and n such that

$$5mn = q \tag{20}$$

and

$$5m^2n^2 = s \tag{21}$$

and

$$m^5 + n^5 = t. \tag{22}$$

Dulaurens claimed that conditions (20) and (22) can always be satisfied because m^5 and n^5 are simply the roots of a quadratic equation in which the sum of the roots is t and the product is $(q/5)^5$. What he ignored for the moment was equation (21) which is consistent with (20) only if $q^2 = 5s$. Not until right at the end of the chapter did Dulaurens add a warning that q and s must be correctly related. In all his worked

[18] Viète 1646, 286–304; 1983, 418–450.

[19] Viète 1646, 305–322; not included in Viète 1983. Viète wrote the equation of degree 45, proposed to him by Adriaan van Roomen, in notation similar to that suggested by Stevin ten years earlier, and as a relationship between proportionals, much as Stevin would have done (perhaps because this was how van Roomen had presented it to him): *Si duorum terminorum prioris ad posteriorem proportio sit, ut* 1① ad 45① −3795③ +9,5634⑤− [...] +945㊶ −45㊸ +1㊺ *deturque terminus posterior, invenire priorem.* [If of two terms, the first to the second is as $1x$ to $45x - 3795x^3 + 9{,}5634x^5 -$ [...] $+ 945x^{41} - 45x^{43} + 1x^{45}$.

[20] Briggs 1633, 3–20.

[21] Since $g = 2r\sin 5\theta$ and $a = 2r\sin\theta$, equation (18) is equivalent to the trigonometric identity $\sin 5\theta = 5\sin\theta - 20\sin^3\theta + 16\sin^5\theta$.

examples they are so, probably because he would have constructed his examples by working backwards from the solution. Thus he was able to show that a root of

$$a^5 - 10a^3 + 20a - 18 = 0,$$

for instance, is

$$a = \sqrt[5]{16} + \sqrt[5]{2}.$$

A comparison of (19) with (18) makes it clear that the fifth degree equations that Dulaurens could solve in this way were simply angle division equations, in which r^2 is replaced by mn and g by $(m^5 + n^5)/m^4n^4$.

Dulaurens applied a similar technique to find equations of degree 7 or 11 which he could solve in a similar way. Thus, for example, he could show that a root of

$$a^{11} - 22a^9 + 176a^7 - 616a^5 + 880a^3 - 352a - 96 = 0$$

is

$$a = \sqrt[11]{64} + \sqrt[11]{32}.$$

Clearly his method applies only to a restricted class of equations in which the coefficients are correctly related. Nevertheless, this was the first breakthrough into solving equations of degree higher than four by an algebraic method.

In an 'Additamentum' at the end of the book, Dulaurens had one further good idea, of a rather different kind. Cardano's method for removing the second term of any equation was by now very well known; Dulaurens saw how a similar technique could in principle be used to remove *any* term from an equation. Given, for instance, the equation

$$a^4 - pa^3 + qa^2 - ra + s = 0,$$

putting $a = e + m$ gives

$$\begin{aligned} e^4 &+ (4m - p)e^3 + (6m^2 - 3pm + q)e^2 \\ &+ (4m^3 - 3pm^2 + 2qm - r)e + (m^4 - pm^3 + qm^2 - rm + s) = 0. \end{aligned}$$

To remove the second term, we need $4m - p = 0$; to remove the third term we must have $6m^2 - 3pm + q = 0$; and so on. (Clearly removing the final term, thereby effectively reducing the degree of the equation, is no easier, in fact exactly the same as, solving the original equation.)

The method suggested by Dulaurens will only remove one chosen term, but it seems that he also began to glimpse the possibility of multiple removal: in his dedicatory preface he claimed that his method could be extended to removing two, or three terms, and that even more would be desirable. 'I know', he added, 'that this will seem a paradox to many, who persuade themselves that everything that men can acquire by human ingenuity is already be found in those things that have been recently written on analysis'.[22] Recall that Harriot had also investigated equations with one or two missing terms back in the 1600s (see page 41), but in a different context. He had been interested in the special relationships between the roots that would result in the disappearance of one or more terms. Dulaurens, on the other hand, was concerned with transformations that would deliberately force such a removal. In other words, Harriot had made observations about existing conditions, whereas Dulaurens hoped to make an active intervention. Dulaurens gave no clues, however, as to how it might be done.

The first short notice of the *Specimina*, only six lines long, appeared in the *Philosophical Transactions of the Royal Society* in December 1667.[23] In other circumstances that might have been all the attention the book ever received. At the end of his text, however, Dulaurens had added and solved a problem concerning line segments in an ellipse, a problem which, he claimed, John Wallis, Savilian Professor at Oxford, had proposed to the mathematicians of Europe.[24] Wallis reacted furiously, denying that he ever had or ever would set such a trivial challenge. At the same time he complained that the *Specimina* was anyway a poor text, derived for the most part from other writers and full of errors. Dulaurens tried to defend himself, explaining that he had received the problem from a friend and that the attribution to Wallis was perhaps a mistake but not a serious one. He asked Wallis, to explain, however, exactly what was wrong with his book. Wallis obliged, and dragged the contents of the *Specimina* vituperatively through the pages of the *Philosophical Transactions* in a lengthy letter published in two parts in August and September.[25] This was Wallis at his worst, intolerant, bullying, and insensitive to the value of other people's mathematics unless it came from within his own circle of friends. With such publicity, however, the *Specimina* was bound to become well known, and one of the people who became particularly interested in Dulaurens' ideas was John Collins.

Collins, an accountant by profession and an early member of the Royal Society, was always keen to discuss the latest mathematical books and ideas with his extensive circle of correspondents. In November 1670 he wrote optimistically to his friend James Gregory, the able young professor of mathematics at St Andrews, about Dulaurens' hope

[22] *hanc methodum sequitur alia multo admirabilior, per quam cuislibet aequationis terminos omnes intermedios auferre licet, et quidem duos, aut tres per ea quae huc usque reperta sunt, verum ad plures quàm tres aufferendos necesse est ut nova reperta dentur, quae generalis hujus methodi usum latius extendant. Scio hoc paradoxum multis virum iri qui sibi persuadent omnia quae humani ingenii viribus acquiri possunt jam ab iis, qui de analysi nuper scripserunt inventa esse, aut facilè ex eorum principiis deduci posse;* [there follows a method much more wonderful than any other, by which all the intermediate terms may be removed from any equation, and indeed two or three by those [methods] so far discovered, but it is necessary to remove more than three so that new discoveries may be found, which greatly extend the use of this general method. I know that this will seem a paradox to many, who persuade themselves that all that men can acquire by human ingenuity is already to be found in those things that have been recently written on analysis, or easily deduced from their principles.] Dulaurens 1667, sig. b.

[23] *Philosophical Transactions*, 2 (1666–67), 580.

[24] The problem originated with one Simon de Montfert (possibly Blaise Pascal), who sent it to the mathematicians of England. It was discussed by Wallis and Brouncker in May 1658, at which time printed versions were in circulation; see Beeley and Scriba, II, letters 159 and 160. Solutions by Christopher Wren and Jonas Moore survive in MS Aubrey 10 in the Bodleian Library, Oxford.

[25] Wallis 1668a; Dulaurens 1668; Wallis 1668b, 1668c.

of finding a method to take away all the middle terms of an equation:[26]

> Dulaurens in his Praeface of his treatise of Algebra promiseth a method whereby to take away all the middle tearmes of any Aequation leaving only the highest and lowest power equall to the Absolute or Homogeneum.

As so often, Collins did not quite understand what was being suggested, for he seems to have thought that the 'highest and lowest' powers could remain, whereas Dulaurens had definitely suggested removing *all* the intermediate powers (*terminos omnes intermedios*), that is, all powers except the highest. Nevertheless, Gregory tested the idea. Just over a year later, in January 1672, he reported some progress but also some severe difficulties:[27]

> a sursolid [fifth-degree] equation, which can be reduced to a pure one, must first ascend to the twentieth potestas [power], not without extraordinary work.

He did not explain this assertion to Collins but his technique survives in two brief manuscripts, eventually published in 1939 in the *James Gregory tercentenary volume*.[28]

As one might expect from Gregory, his method was both innovative and thoughtful. It goes like this. Given an equation $x^3 + q^2x + r^3 = 0$, first make the substitution $x = z + v$ to obtain

$$v^3 + 3zv^2 + (q^2 + 3z^2)v + (z^3 + q^2z + r^3) = 0.$$

Now multiply this by the cubic expression $v^3 + av^2 + b^2v + c^3$ to obtain an equation of degree 6 in v. Gregory's idea was to choose a, b, c, and z in such a way that the coefficients of v^5, v^4, v^2, and v would vanish, leaving a quadratic equation in v^3. To achieve this the following four equations must be simultaneously satisfied:

$$\begin{aligned} 3z + a &= 0 \quad \text{(coefficient of } v^5\text{)},\\ q^2 + 3z^2 + 3za + b^2 &= 0 \quad \text{(coefficient of } v^4\text{)},\\ z^3 + q^2za + r^3a + b^2q^2 + 3z^2b^2 + 3c^3z &= 0 \quad \text{(coefficient of } v^2\text{)},\\ b^2z^3 + b^2q^2z + b^2r^3 + c^3q^2 + 3c^3z^2 &= 0 \quad \text{(coefficient of } v\text{)}. \end{aligned}$$

Gregory eliminated a, b^2, and c^3 in turn to arrive at a cubic equation in z^2:

$$27z^6 - 27r^3z^3 - q^6 = 0.$$

This equation is solvable and so in principle, Gregory had achieved what he wanted. He did not, or at least not on this sheet of paper, follow through the rest of the working, which would have entailed first solving for z, then solving the equation of degree 6 for v (by now reduced to a quadratic in v^3), then calculating values of $v + z$. He did

[26] Collins to Gregory, 1 November 1670, in Gregory 1939, 111.

[27] Gregory to Collins, 17 January 1672, in Gregory 1939, 210–212, and Rigaud 1841, II, 229–231.

[28] Gregory 1939, 382–390.

not need to: his aim was not to solve cubic equations, which had already been done, but to discover a method that might work for equations of higher degree.

The method extends fairly easily to equations of degree 4 of the form $x^4 + q^2x^2 + r^3x + s^4 = 0$. Here Gregory made the same substitution $x = z + v$ and then multiplied by $v^2 + av + b^2$ to arrive again at an equation of degree 6. This time he eliminated the coefficients of v^5, v^3, and v, leaving a cubic in v^2, which is solvable. It was natural to consider next if a similar method could be applied to equations of degree 5. After making the usual substitution $x = z + v$, Gregory bravely multiplied his equation by an expression of degree 15 to arrive at an equation of degree 20. This time he needed to eliminate all the coefficients except those belonging to the powers 20, 15, 10, 5, and 0; in other words to reduce the equation of degree 20 to a quartic in v^5. This, however, meant eliminating 16 unknowns from 16 equations. Gregory saw no reason to suppose that this could not be done by someone who was not afraid of the labour.

Gregory's letter of January 1672 ended with tantalizing hints of further results:

> I could send you several general notions of all equations, which, for what I know, are yet untouched by any; but I am afraid they should hardly be so pleasing to you, as it were troublesome to me, to seek them out, and transcribe them; I being now upon another study.

Gregory may have been upon other studies but Collins was not easily deflected. A further hint from Gregory in the spring of 1675 that he had had some success in 'reduction of equations and finding all the roots' set Collins on the trail again.[29] In reply, Gregory once again commented on the removal of terms, arguing that it was easy enough to find special cases where the disappearance of one term would entail the disappearance of others. In the general case, however, he asserted that removal of terms could only lead to equations of yet higher degree:[30]

> It is easy to constitute equations so that either two, three, &c., or all the intermediate terms, may easily go off; but to take off even two intermediate terms in an arbitrary equation, without elevating it, is absolutely impossible. By elevating it I can take away all the intermediate terms myself, which (so far as I know) the world is yet ignorant of.

Collins could only have been disappointed by such a reply. By the time it reached him, however, he had found another potential ally: at the beginning of May 1675 Ehrenfried Walter von Tschirnhaus arrived in London. Almost immediately Collins tried to engage his help on Dulaurens' conjecture.

Tschirnhaus came from a landowning family from the region that is now the meeting point of Germany, Poland, and the Czech Republic. Not needing to work for his living, he spent his early adult years studying and travelling in the Netherlands, England, France, and Italy, where he met and befriended some of the foremost mathematicians and scientists of the day. As a student in Leiden, Tschirnhaus was taught mathematics

[29] Collins to Gregory, 1 May 1675, in Gregory 1939, 298–302.

[30] Gregory to Collins, 26 May 1675, in Gregory 1939, 303, and Rigaud 1841, II, 260.

by Pieter van Schooten, the younger half brother of Frans van Schooten, editor of the *Geometria*, and he became a fervent supporter of Cartesian methods. At the age of 24, he came to London where over several weeks he met many of the leading members of the Royal Society.

Collins wrote to Henry Oldenburg, secretary of the Society, on 25 May 1675 begging him to put the following problem to Tschirnhaus:[31]

> Be pleased to intreate the learned and worthy Mr Tschirnhaus to make a Construction by a Circle for finding a roote of
>
> $$aaa - 3aa + 3a - 1 = N.$$

It is difficult to see exactly what Collins meant by this. He almost certainly knew that an angle trisection equation took the form $x^3 = px \pm q$. Did he simply want Tschirnhaus to remove the term in aa, 'leaving only the highest and lowest power' (as he had suggested in 1670)? In this case the usual technique for removing the square term removes the linear term as well, which perhaps unsettled Collins. However, Oldenburg, as requested, passed the problem to Tschirnhaus in the form that Collins had posed it, and Tschirnhaus wrote to Oldenburg the next day to say:[32]

> You will remember those things you mentioned to me yesterday. How a cubic equation … might be resolved by means of a circle.

Whatever Collins and Tschirnhaus meant by resolving an equation 'by means of a circle', it seems that Tschirnhaus took on board the idea of eliminating intermediate terms. In August 1675 Collins wrote to Gregory in some excitement that Tschirnhaus was '(excepting your selfe and Mr Newton) […] the most knowing algebraist in Europe' and that he knew how to remove two terms from certain quartic equations. Unfortunately these were just special cases, as Gregory was quick to point out: if in the equation $x^4 - px^3 + qx^2 - rx + s = 0$, for instance, it happens that $p^3 + 8r = 4pq$ (or as Gregory wrote it: $\frac{p^2}{4} + \frac{2r}{p} = q$) then removal of the cube term will automatically entail the removal of the linear term as well.[33] Gregory had already found such cases himself, as he had mentioned to Collins back in May, but Collins had failed to understand him.

Tschirnhaus took no more heed of Gregory's warnings about special cases than Collins had done. In August 1676 he wrote again to Oldenburg:[34]

> As for the resolution of equations by the removal of all intermediate terms, this is assuredly easy.

[31] Collins to Oldenburg, 25 May 1675, in Oldenburg 1965–86, XI, 323–324.

[32] Tschirnhaus to Oldenburg, July 1675, in Oldenburg 1965–86, XI, 409–411.

[33] Collins to Gregory, 3 August 1675, in Gregory 1939, 314–320; Gregory to Collins, 20 August 1675, in Gregory 1939, 324–326, and Rigaud 1841, II, 269–272.

[34] Tschirnhaus to Oldenburg, 22 August 1676, in Oldenburg 1965–86, XIII, 53–56.

The examples he offered to Oldenburg, however, were precisely the kind that Gregory had dismissed as trivial, where if one term was made to vanish a second term would vanish also. He was still a long way from having a general method, but remained undaunted:

> However, if this is to be done with an arbitrary equation ... still I do not see the impossibility.

By April 1677 Tschirnhaus had finally come up with an idea that might work, as he explained in a letter to Leibniz:[35]

> If now we want to remove two terms in any equation, it must certainly be supposed that $xx = ax + b + y$, if three $x^3 = axx + bx + c + y$, if four $x^4 = ax^3 + bxx + cx + d + y$, and so on as far as you like, regardless of the demonstration to the contrary that Gregory has put forward, according to what Oldenburg has written.

Tschirnhaus's idea began from the well known method for removing the second term from any equation in x using the substitution $x = a + y$ for a suitable value of a. Now he was arguing that it should be possible to remove two intermediate terms by means of a substitution $xx = ax + b + y$ with suitable values of a and b; or three intermediate terms by means of the substitution $x^3 = axx + bx + c + y$; and so on. Testing his idea on a cubic equation of the form $x^3 - qx - r = 0$, using the substitution $xx = ax + b + y$, he found that by putting $b = \frac{2q}{3}$ and $a = \frac{3r}{2q} \pm \sqrt{\frac{9rr}{4qq} - \frac{q}{3}}$ he obtained a simple equation for y (simple in the sense that it requires only the extraction of a cube root):

$$y^3 = 4rr - \frac{27r^4}{2q^3} - \frac{8q^3}{27} + \overline{\frac{4qr}{3} - \frac{9r^3}{qq}}\sqrt{\frac{9rr}{4qq} - \frac{q}{3}}$$

(where the overline indicates bracketing of terms). Thus he could easily find a value for y and therefore for x.

Like Gregory, Leibniz expressed caution:[36]

> I do not think this can succeed in equations of higher degree except in special cases ...

In the face of such doubts from both Gregory and Leibniz, a more modest or sensitive person might have withdrawn in silence, but Tschirnhaus was not to be deflected. In 1683, he published his idea in the *Acta eruditorum*, setting out his transformations just

[35] *Si jam velimus duos terminos in quacunque aequationue auferre, supponendum saltem* $xx = ax+b+y$, *si tres* $x^3 = axx + bx + c + y$, *si quatuor* $x^4 = ax^3 + bxx + cx + d + y$, *atque sic in infinitum, non obstante demonstratione qua contrarium evincebat Gregorius, prout scribit Oldenburgerus*. Tschirnhaus to Leibniz, 17 April 1677, in Leibniz 1976 (3), II, 65–68.

[36] *non puto succedere posse in altioribus nisi quoad casus speciales*. Leibniz to Tschirnhaus, [late December 1679], in Leibniz 1976 (3), II, 924–925.

as he had communicated them to Leibniz six years earlier: two intermediate terms were to be removed by means of a substitution $xx = bx + y + a$ with suitable values of a and b; three intermediate terms by means of the substitution $x^3 = cxx + bx + y + a$; and so on. In this way, he claimed, it should be possible to reduce any polynomial equation of degree n to the simple and easily solvable form $y^n - N = 0$. By now he was able to show that his solution for a cubic equation was equivalent to Cardano's, though the algebraic manipulations were not easy. Ignoring Leibniz's warning, he also ventured into equations of higher degree. For equations of fourth, fifth, or sixth degree, with their second term already removed, he found values of a and b that would serve to remove a further term (of degree 2, 3, 4, respectively) but did not complete the working. If he had, or had attempted to remove more terms, he would have found the technicalities exceedingly difficult.[37] Unlike Gregory, however, Tschirnhaus did not pursue his method far enough to see where the problems lay. Thus although the intention behind his method was clear enough, its applicability remained largely untested.

During the late 1670s, equation solving seems to have captured Tschirnhaus's interest, because the removal of intermediate terms was not the only method he tried. After he left London in the autumn of 1675 he explored the subject with Leibniz in Paris, and they continued to discuss it in their correspondence after Leibniz moved to Hannover a year later. Their letters remained unpublished until the late nineteenth century so that, as with Gregory, their ideas had no direct historical influence, but they are nevertheless of considerable interest in relation to themes to be explored in Part II of this book.[38]

In April 1677, as we saw above, Tschirnhaus sent Leibniz his suggestions for removing intermediate terms. By the end of that year he informed Leibniz that by now he had three methods of solving equations.[39] The first was based on Hudde's idea that one could separate the sought quantity x into two parts, thus $x = a + b$. Tschirnhaus suggested that one might separate x into more parts, for example, $x = a+b+c$. As a preliminary he tabulated powers of $a+b, a+b+c, a+b+c+d$ and of $ab+ac+bc$, and so on, and noted some of the symmetries in these expressions but his exposition fails to make clear how he intended to use any of this to solve equations. His second method consisted of trying out expressions for x involving surds, like $x = \sqrt{a} + \sqrt{b}$ or $x = \sqrt[3]{a} + b$ or $x = \sqrt{a + \sqrt{b} + \sqrt{c}}$, and examining the equations one arrived at by liberating such expressions from their radical signs. Tschirnhaus could easily show, for example, that putting $x = \sqrt[3]{a} + \sqrt[3]{b}$ led to the equation

$$(x^3 - a - b)^3 = 27x^3ab,$$

or

$$x^3 - a - b = 3x\sqrt[3]{ab}.$$

The presupposed solution $x = \sqrt[3]{a} + \sqrt[3]{b}$ agrees with the solution found by Cardano's rule for this equation, which seemed to confirm for Tschirnhaus that his idea was a

[37] It was not until 1786 that the Swedish mathematician Erland Samuel Bring succeeded in using Tschirnhaus transformations to remove the second, third, and fourth powers from a quintic equation.

[38] Leibniz 1899 and Leibniz 1976 (3), II.

[39] Tschirnhaus to Leibniz, late November 1677, in Leibniz 1976 (3), II, 285–286.

good one. The third method was the removal of terms, which he had already sent to Leibniz earlier.

Tschirnhaus expanded on his first method at very great length the following April.[40] Leibniz's reply was robust.[41] He pointed out that he himself had already shown Tschirnhaus the second method when they were in Paris, and that Tschirnhaus had scorned it then but now seemed to have arrived at the same idea himself. As for the first method, they had discussed that too in Paris, at which time Leibniz had already discovered its shortcomings. One of them was that the method would take an equation of degree 4 to one of degree 12, and an equation of degree 5 to one of degree 20 (as Gregory had also found), whereas Tschirnhaus seemed to believe he could *reduce* the degree of an equation. (Leibniz, however, appears also to have believed that equations of degree 8, 9, or 10 could be reduced to seventh-degree equations.)[42] In short, said Leibniz, although the method appeared to offer a way in to the problem, it offered no way out (except for cubic equations) as Tschirnhaus would surely discover if he were to calculate even one example.

Despite this sharp rebuttal, the correspondence between Tschirnhaus and Leibniz continued during the rest of 1678 and 1679, though several of the letters are now missing.[43] Tschirnhaus was not easily persuaded that his suggestions were futile, but in December 1679 Leibniz, who was probably growing thoroughly weary of the discussion, sent concise and dismissive responses to all three methods.[44] The first, he said once more, was likely to lead to a situation from which there was no way out;[45] for the second, the labour of calculation would need to be immense;[46] and the third could not succeed except in special cases.[47] This last remark in fact applied to all three of Tschirnhaus's methods. The ideas Leibniz had pursued in Paris and that Tschirnhaus later took up were not absurd in themselves: in the examination of symmetric functions, and consideration of roots as sums of radicals, Leibniz came very close to ideas that were pursued by Euler and Bezout later, but was defeated by the complexity of the calculations. Tschirnhaus did not go far enough even to be defeated.

As for Gregory, the final round of correspondence between him and Collins in 1675 shows that he, like Leibniz, had penetrated the matter quite deeply. In May 1675 he wrote:[48]

> I have now abundantly satisfied myself in these thing[s I] was searching after in analytics, which are a[ll about] reduction and solution of equations. It is possible that [I flatter] myself too much, when I think them of some value, [and] therefore am sufficiently inclined to know others' thoughts,

[40]Tschirnhaus to Leibniz, 10 April 1678, in Leibniz 1976 (3), II, 369–381.

[41]Leibniz to Tschirnhaus, [May/June 1678], in Leibniz 1976 (3), II, 422–445.

[42]Leibniz to Tschirnhaus, [May/June 1678], in Leibniz 1976 (3), II, 423, 431.

[43]See Leibniz 1976 (3), II, letters 192, 208, 301, 309, 362.

[44]Leibniz to Tschirnhaus, [late December 1679], in Leibniz 1976 (3), II, 923–925.

[45]*deprehendi istam methodum non posse ad exitum perducere.*

[46]*sed calculo opus esset immenso.*

[47]*non puto succedere posse in altioribus nisi quoad casus speciales.*

[48]Gregory to Collins, 26 May 1675, in Gregory 1939, 302–305, and Rigaud 1841, II, 259–266.

> both (as ye say) as to the quid and quomodo of them; but that I have no ground to expect, till time and leisure suffer me to publish them.

As we saw above, however, Gregory's discoveries had led him into tortuous calculation. When Collins told him in August that Tschirnhaus promised a general method, Gregory hoped it would be more concise than his own, for if it were not, he wrote, 'I question if a twelvemonth shall serve for to calculate the canons for the equations of the first ten dimensions'.[49] For cubic and quartic equations, Gregory claimed, his own method was the most efficient he had yet seen, but 'the labour increases at a strange rate as the dimensions augment'. He made an offer, however, that Collins could hardly refuse: 'If any would undertake to calculate the canons, I would willingly communicate the method with its demonstration'. Collins wrote back immediately to suggest that his friend Michael Dary might be the person to undertake such calculations; Dary, he said, had recently 'improved himself in Algebra' by 'reading Kinckhuysen at the farthing Office, where I got him to attend, during my absence in the mornings'.[50] Gregory, knowing the scale of the problem far better than Collins, can hardly have been encouraged by this account of Dary's new skills, and he retreated, saying he was 'extremely apprehensive that Mr Dary hath not patience enough for such tedious work'.[51] Collins wrote once more, on 19 October, begging Gregory to make a clear proposal concerning the calculations he needed, a proposal which he, Collins, would then put to the Royal Society.[52] He was too late: that same month Gregory suffered a stroke and died a few days later aged 36. To our very great loss he never did have 'time and leisure' to publish what he knew.

Curves and limits, 1669–1691

The idea of representing the values of a polynomial by sketching a series of ordinates (in modern terms the 'y values') was perhaps first used by Isaac Newton in 1664–65 in his researches into the binomial theorem, where he drew families of curves, for instance, $y = (1 - xx)^{\frac{n}{2}}$ for values of x between 0 and 1 and $n = 0, 1, 2, 3, 4, 5, 6$, and $y = x^n$ for $n = -1, 0, 1, 2$. Other sketches, of $y = \frac{1}{x^2}$ or $y = x^2 + x^{\frac{3}{2}}$, for example, appear in his 'De analysi' written and shown to Isaac Barrow five years later.[53] It is perhaps no coincidence, then, that the first published treatment of curve-sketching appeared in the final chapter of Barrow's *Lectiones geometricae* in 1670. Barrow began the chapter by asserting that Viète had explained the nature of equations by the proportions (*analogia*) of their terms, while Descartes had done so more lucidly in terms of multiplication of factors, but that he, Barrow, would now offer a different kind of description, by use of curved lines, which would present the matter 'to the eye'.[54] Although Barrow used

[49]Gregory to Collins, 20 August 1675, in Gregory 1939, 324–326, and Rigaud 1841, II, 269–272.

[50]Collins to Gregory, 4 September 1675, in Gregory 1939, 327–328.

[51]Gregory to Collins, 11 September 1675, in Gregory 1939, 328–330, and Rigaud 1841, II, 272–274.

[52]Collins to Gregory, 19 October 1675, in Gregory 1939, 337–344, and Rigaud 1841, II, 277–281.

[53]Newton 1967–81, I, 104, 112, 122, 123; see also II, 208–210.

[54]Aequationum *naturam è terminorum* analogia *exposuit* Vieta; *illam ex eorum in se ductu dilucidius explicuit* Cartesius. *Eam ego jam è linearum singulis appropriatarum descriptione conabor aliquatenus*

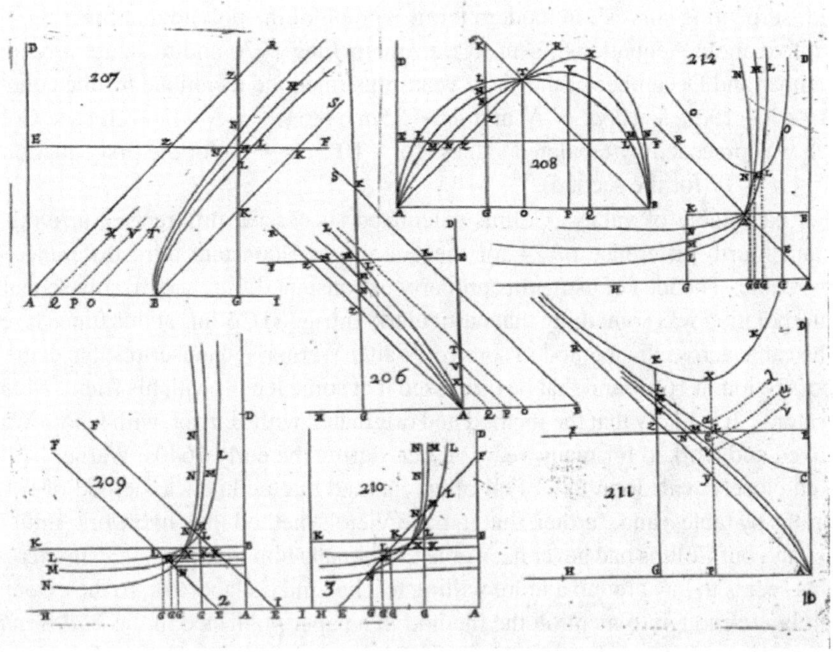

A new understanding of equations (3): 'serpentine curves', from Barrow's *Lectiones* (1670).

Cartesian superscript notation for powers (apart from squares) he retained Harriot's use of a for the unknown quantity, and also maintained strict homogeneity. He did not use co-ordinate axes in the modern sense, but erected ordinates along a base line to show the value of the polynomial for each value of a. His curves appear in groups because, for instance, he plotted $b-a=n$, $ba-aa=nn$, $baa-a^3=n^3$, and $ba^3-a^4=n^4$ all on the same diagram, on a base line AB representing the length b (and therefore, in modern terms, from $a=0$ to $a=b$). Barrow's representations showed clearly what had long been known about such equations, namely, that for appropriate values of n^2, n^3 and so on (and apart from the trivial case $b-a=n$) they each have two real positive solutions.[55] Other information, such as the position and value of a maximum, or the relative slope of a tangent, could also be seen from his sketches.

By June 1670, as the *Lectiones geometricae* went to press, Barrow's friend and correspondent John Collins (another recipient of Newton's *De analysi*) was also interested in representing values of polynomials by a series of ordinates to produce what he

enucleatam dare; qui sanè modus rem praesertim elucidare videtur, ac ob oculos ponere, agedum. [Viète expounded the nature of equations from the proportions of their terms; Descartes explained it more clearly from parts of them multiplied together. I will now present it by description of one appropriate line and will endeavour as far as possible to give something straightforward, which seems to elucidate the matter and put it to the eye in a particularly reasonable way]. Barrow 1670, 131.

[55]Barrow 1670, 133–135.

called 'serpentine curves', in modern terms a graph of the polynomial function.[56] His 'Narrative about Aequations', sent to Barrow in June 1670 and to James Gregory in November and December of the same year, illustrated the technique for the equations $a^4 - 4a^3 - 19aa + 106a = N$ and $a^3 - 15aa + 54a = N$. In each case Collins' sketch was preceded by tabulated values (for $-10 \leq a \leq 10$ for the first equation and for $1 \leq a \leq 17$ for the second).[57]

For each table of values, Collins calculated successive differences, arriving at a constant fourth difference of 24 for the quartic, and constant third difference of 6 for the cubic. The idea of using this property of constant differences to solve equations by interpolation was something that particularly intrigued Collins at this time. It seems that he came across the method in some of Walter Warner's manuscripts that came into his possession in 1667, and that he discussed it at some length with his friend Nicolaus Mercator.[58] It is likely that the method had originated with Harriot, with whom Warner had lived and worked for many years. Later, during the early 1640s, Warner had also worked closely with John Pell. Pell often claimed that he knew a method of solving equations by tables, and, further, that it made Viète's method look like work 'unfit for a Christian', but Collins had never been able to persuade him to explain it.[59] In Mercator, Collins seems to have found a more willing teacher and collaborator, to the extent that he felt brave enough to mention the method in a paper published in the *Philosophical Transactions* in 1669 and in his 'Narrative about equations' in 1670. Unfortunately, Collins was rarely able to explain any mathematical idea comprehensibly, and his writings give no more than hints of the method without any of the essential details.

A further example of a curve corresponding to an equation, this time $a^3 - 48a = N$, appeared in a letter written by Collins and published posthumously in the *Philosophical Transactions* in 1684. The description of the curve appears at the beginning of the piece, the rest of which is a meandering ramble through Collins' muddled knowledge of equations. He did touch on one significant idea though, and one which probably came to him through his study of 'serpentine curves', namely, that the roots of an equation (provided they are real and distinct) are separated by local maxima or minima. Collins described the maxima and minima as the 'dioristick limits', that is the 'defining' or 'distinguishing' limits, because they separate or distinguish between the real roots of the equation. Further, he knew that the maxima and minima can be found by solving an equation of lower degree, as explained by Hudde. Collins described Hudde's method as follows:[60]

> Now for instance (according to *Huddens* method) in a biquadratick aequation, you must multiply all the terms beginning with the highest, and so in order by 4, 3, 2, 1, and the last term or Resolvend by 0. Whereby it is de-

[56] See Collins to Gregory, 1 November 1670, in Gregory 1939, 109–118; Collins to Gregory, 25 March 1671, in Rigaud 1841, II, 219.

[57] Collins 1670.

[58] See Beery and Stedall 2009.

[59] Rigaud 1841, I, 248.

[60] Collins 1684, 578.

> stroyed, and you come to a cubick Aequation, [...] the roots whereof being found, and as roots having Resolvends raised thereto in the biquadratick Aequation, are the dioristick Limits thereof.

Now Collins speculated as to what might happen if the method were repeated:

> And if this easy method were known, we may come down the Ladder to the bottom, and fall into irrational quantities, and ascend again. Against which assymetry, an Aequation might be assumed low, as a rational quadratick, and thence a cubick Aequation formed, whose limits should be found by aid of the quadratic Aequation, and out of that cubick a Biquadratick Aequation, whose limits should be found by the aid of that cubick Aequation, &c.

In other words, the roots of each lower degree equation are limits, or bounds, for the roots of the next equation up.

This was an idea that was to be expressed more precisely just a few years later by Michel Rolle in his *Traité d'algebre*, published in 1690 in Paris.[61] Rolle explained that one should first prepare the equation to be solved, which let us suppose is in x, so that all its (real) roots are positive, and described how this should be done. He claimed that an upper bound for the roots of a polynomial can be found by dividing the absolute value of the lowest negative coefficient by the coefficient of the highest term and adding 1. If this upper bound is B then the transformation $y = B - x$ leads to a new equation, in y, in which all the roots are positive.[62]

Next Rolle instructed that one should form the 'cascade' of the equation by multiplying each term by the number of its degree and then dividing by the unknown. One of his own examples was the equation $y^3 - 57yy + 936y - 3780 = 0$, whose successive cascades are[63]

$$\begin{aligned} y^3 - 57yy + 936y - 3780 &= 0, \\ 3yy - 114y + 936 &= 0, \\ 3y - 57 &= 0. \end{aligned}$$

[61] For a detailed account of Rolle and his method see Barrow-Green 2009.

[62] Rolle gave this rule for an upper bound without proof: *On prendra parmi les termes negatifs de l'égalité, celuy qui a le plus grand nombre connu; on effacera le signe & l'inconnuë de ce terme, on divisera le resultat par le nombre connu du premier terme, & au quotient on ajoûtera l'unité, ou un nombre positif plus grand que l'unité. De la somme qui en viendra on ostera une nouvelle inconnuë, & substituant le reste au lieu de l'inconnuë dans l'égalité proposé, la substitution donnera une autre égalité dont les signes seront alternatifs.* [One selects from the negative terms of the equation that with the greatest coefficient; one ignores the sign and the unknown in this term, one divides the result by the coefficient of the first term, and to the quotient one adds unity, or a positive number greater than unity. From the sums that arises one takes a new unknown and substitutes the remainder in place of the unknown in the proposed equations; the substitution will give another equation in which the signs alternate.] Rolle 1690, 120. For a proof that the rule gives an upper bound for the roots in the case of cubics see Reyneau 1708, I, 93–96, and Maclaurin 1748, 172–174.

[63] Rolle 1690, 127–128. Rolle wrote the equations in the opposite order to that shown here. It is not clear why he did not divide through by 3 in the quadratic and linear equations.

The equation of least degree gives $y = 19$, which is therefore an intermediate limit (Rolle called it a *hypothese moyenne*) for the roots of the quadratic just above it. Outer limits (*hypotheses extremes*) are 0 (from the way the equation has been prepared) and $114/3 + 1 = 39$ (by the rule given above). A set of limits for the quadratic is therefore 0, 19, 39. Now Rolle assumed something that intuitively presents no difficulty: that on either side of a real root a polynomial will take alternatively positive and negative values.[64] Thus, by interval bisection, he was able to 'close in' on the roots, and found that they are 12 (between 0 and 19) and 26 (between 19 and 39). Thus we now have the limits 0, 12, 26, 3781 for the original cubic, and its roots turn out to be 6, 21, and 30.

Rolle's process seems to be what Collins had in mind when he claimed that 'we may come down the Ladder to the bottom [...] and ascend again'. Collins' 'serpentine curves' for equations with real distinct roots demonstrate visually that the turning points of the curve alternate with the roots. Rolle too may have been guided by some such picture, but no diagrams appear in his text. He certainly never spoke of 'differentiating' in the modern sense, and indeed had little time for the new 'analysis of the infinitely small'.[65] Rather, his procedure was the one that Hudde had suggested: term by term multiplication by a general arithmetic progression. Using the power of each term as the multiplier is particularly convenient because it causes the constant term to disappear, but it is not essential. In a *Démonstration* published a year after the *Traité d'algebre*, Rolle proved purely algebraically that if a and b are consecutive roots of a polynomial, the first cascade, or derived polynomial, is negative at a and positive at b, or vice versa, and therefore has a root between a and b. His proof was conceptually similar to that used by Hudde in 1658 to show that a double root of an equation is also a root of its derived equation. Rolle argued that if $(x - a)(x - b)$, that is, $xx - (a + b)x + ab$ is multiplied term by term by any arithmetic progression, say $y + 2v$, $y + v$, y, then we have the cascade

$$(y + 2v)xx - (y + v)(a + b)x + yab,$$

or

$$y(x - a)(x - b) + v(2xx - (a + b)x).$$

When $x = a$ the value of the cascade is $va(a - b)$ and when $x = b$ it is $vb(b - a)$, and therefore of opposite sign. The existence of a further factor such as $(x - c)$ makes no difference to the argument because, since a and b are consecutive roots, $(a - c)$ and $(b - c)$ will always have the same sign.

Rolle's theorem, as it came to be called, that between two real roots of an equation there is always a root of the derived equation, was later generalized to any differentiable function and became one of the cornerstones of Analysis. It emerged first, however, as an algebraic theorem, in the context of solving polynomial equations.

---

[64] *Lors qu'il y a des racines effectives dans un cascade, les hypotheses de cette cascade donnent alternativement l'une + & l'autre −.* [When there are real roots in a cascade, the limits of this cascade give alternatively negative and positive values.] Rolle 1690, 128.

[65] Barrow-Green 2009.

Newton's *Arithmetica universalis*, 1707

Just as Viète's *Tractatus duo*, published in 1615 can be seen as the last word on the theory of equations for the sixteenth century, so Newton's *Arithmetica universalis*, published in 1707, stands as the culmination of the theory for the seventeenth century. Written piecemeal over many years, it began as a set of notes and elucidations on the Cartesian theory of equations as it was still being worked out in the 1660s, but ended up including some of Newton's own ideas from the 1680s. The book thus covers much the same time span as the present chapter, and incorporates many of the ideas that have already been discussed, but as sifted and compiled by Newton.

In the late 1660s Nicolaus Mercator made a Latin translation of Gerard Kinckhuysen's *Algebra, ofte stel-konst*, published in 1661. He did so probably at the request of John Collins, who was always looking for new mathematical texts for English readers. Kinckhuysen's *Algebra* was the first elementary textbook to take up ideas from the first volume of the *Geometria* published two years earlier. In addition to the usual procedures for manipulating and simplifying equations (clearing fractions, removing terms, and so on) Kinckhuysen taught Descartes' rule of signs, Cardano's rule for cubics, Descartes' method for quartics, and Hudde's rule for discovering double roots. All of this was considerably more advanced than anything yet published in England. Nevertheless, Collins felt that the text would benefit from some additional notes and clarifications, and in December 1669 asked Newton, whom he had only recently met, to provide them. Newton worked on the notes in the course of the following year, but in the end Mercator's translation was never printed and Newton's notes remained unpublished.[66]

Under the Lucasian statutes Newton was required to deposit ten lectures a year in the University Library, and in 1684 he submitted a set of notes, which, according to dates inserted in the margins, represented lectures delivered from 1673 to 1683. The supposed lecture notes include his annotations on Kinckhuysen's *Algebra*, together with a great many new examples. Newton may indeed have taught some of this material, but an entire lecture series of the kind indicated by his notes almost certainly never existed: the material is cumulative, and students who began later than the first year would have found themselves impossibly bewildered by a series of difficult examples for which they had received no training. These, however, were the notes that were edited and published in 1707 by William Whiston as the *Arithmetica universalis*. In its overall structure and content the *Arithmetica universalis* retained many of the features of Kinckhuysen's *Algebra* from half a century earlier. But Newton had also inserted into his supposed lecture notes some new discoveries of his own. As was typical of Newton (and, as we have seen, of many of his predecessors also) these were presented through rules and worked examples without either proof or explanation, and therefore

[66] Mercator's translation and Newton's annotations are inserted into a copy of Kinckhuysen's *Algebra* now held in the Bodleian Library, Oxford (Savile G.20). For full transcripts of both see Newton 1967–81, II, 295–447. For discussion of the failed plans for publication see Scriba 1964 and Whitesides's account in Newton 1967–81, II, 277–291.

required a considerable amount of expository work by others later.

The first of Newton's innovations came in a section near the beginning of the book headed 'De inventio divisorum' ('On finding divisors'), in which he offered several examples of a new method for finding divisors of polynomials.[67] Here is his method illustrated for $x^3 - xx - 10x + 6$. First evaluate the polynomial when $x = 1, 0, -1$ to give -4, 6, 14, respectively. Next write down all the divisors of 4, 6, and 14, as shown in the table below. Then search amongst the divisors (any of which may be regarded as positive or negative, though Newton did not actually say so) for arithmetic progressions with a common difference of 1. In this case we may take 4, 3, 2, from the first, second, and third row, respectively.

1	4	1.	2.	4.		+4.
0	6	1.	2.	3.	6	+3.
−1	14	1.	2.	7.	14	+2.

The fact that 3 appears in the row starting with 0 suggests that $x + 3$ should be tested as a divisor (because it takes the values 4, 3, 2 when x is given the values $1, 0, -1$). And indeed it is the case that $x^3 - xx - 10x + 6 = (x + 3)(xx - 4x + 2)$. Newton gave similar but rather more complicated rules for finding quadratic divisors, or divisors of equations with literal coefficients, but without any explanation of the underlying principles.

Another new topic that appeared quite early in the *Arithmetica universalis* was what Newton called 'De duabus pluribusve aequationibus in unam transformandis ut incognitae quantitates exterminentur' ('On transforming two or more equations into one so that unknown quantities are eliminated').[68] Here Newton showed how to use one equation to eliminate a given quantity from a second equation. He also gave four rules, or rather conditions, that must hold if an unknown quantity is to be eliminated from two polynomials. If, for example, $axx + bx + c = 0$ and $fxx + gx + h = 0$ are to hold simultaneously, then, according to Newton, it must be the case that

$$(ah - bg - 2cf)ah + (bh - cg)bf + (agg + cff)c = 0. \tag{23}$$

Newton did not explain how to obtain this equation but he showed by example how to use it. Thus if $xx + 5x - 3yy = 0$ and at the same time $3xx - 2xy + 4 = 0$, then y must satisfy

$$316 + 40y + 72yy - 90y^3 - 69y^4 = 0.$$

Newton's equation (23) came to be known as the 'elimination equation' for the two equations in question.

Newton's remaining discoveries on equations appear only towards the end of the book. Following his statement of Descartes' rule of signs, he gave another and completely new rule, for finding the number of 'impossible' (imaginary) roots.[69] This

[67] Newton 1707, 42–51; 1720, 38–47.

[68] Newton 1707, 69–76; 1720, 60–67.

[69] Newton 1707, 242–245; 1720, 197–200.

was to cause considerable perplexity to his readers because, in his usual way, Newton presented worked examples without any explanation. We will do the same here, and leave until Chapter 4 the attempts of later writers to justify the procedure. Thus, take Newton's equation

$$x^5 - 4x^4 + 4x^3 - 2xx - 5x - 4 = 0.$$

Now take the fractions $\frac{1}{5}, \frac{2}{4}, \frac{3}{3}, \frac{4}{2}, \frac{5}{1}$, divide each by the one that follows, and write the results above the inner terms of the equation, thus

$$\begin{array}{ccccccc} & \frac{2}{5} & \frac{1}{2} & \frac{1}{2} & \frac{2}{5} & & \\ x^5 & -4x^4 & +4x^3 & -2xx & -5x & -4 & = 0. \\ + & + & - & + & + & + & \end{array}$$

The sign below a term is then $+$ or $-$ according to whether the square of the term, multiplied by its overhead fraction, is greater or less than the product of the terms on either side. In this case, for instance, $\frac{2}{5}(4x^4)^2 = \frac{32}{5}x^8 > x^5 \times 4x^3 = 4x^8$ so a $+$ sign is placed below $-4x^4$. A $+$ sign is also placed under each of the end terms. Each change of sign from $+$ to $-$ or $-$ to $+$ then indicates the existence of an 'impossible' root. Newton did not explain what to do when the comparison yields an equality. This happens here for $\frac{1}{2}(4x^3)^2 = 8x^6 = (-4x^4)\times(-2xx)$, where Newton silently inserted a $-$ sign.

After his rule for 'impossible' roots, Newton turned to the transformation of equations ('De transmutationibus aequationum') and gave the usual rules for augmenting or diminishing the roots. Then he went on to discuss the composition of the coefficients from the roots, and gave the following rules, exactly equivalent to those given by Girard in 1629 but in more easily memorable form.[70] Suppose $-p$, $-q$, $-r$, $-s$, ... are the coefficients of an equation from the second highest term downwards, and that a is the sum of the roots, b the sum of their squares, c the sum of their cubes, and so on. Then, according to Newton,

$$\begin{aligned} a &= p, \\ b &= pa + 2q, \\ c &= pb + qa + 3r, \\ d &= pc + qb + ra + 4s, \\ e &= pd + qc + rb + sa + 5t, \\ f &= pe + qd + rc + sb + ta + 6v. \end{aligned}$$

If all the roots are real, Newton argued, then $\sqrt{b}$, $\sqrt[4]{d}$, $\sqrt[6]{f}$, ... give increasingly good estimates for the root with the largest absolute value. He was able to derive other rules,

[70] Newton 1707, 251–252; 1720, 205–206.

too, for equations where all but two of the roots are negative, for instance, but as the estimates become tighter the calculations become more complicated.[71]

Newton therefore proposed another method for finding an upper bound for the roots. He described it as multiplication by an arithmetic progression.[72] Take, for example, the equation

$$x^5 - 2x^4 - 10x^3 + 30xx + 63x - 120 = 0. \tag{24}$$

Multiply the left hand side term by term by the progression 5, 4, 3, 2, 1, 0, and divide by x to give

$$5x^4 - 8x^3 - 30xx + 60x + 63.$$

Continuing in a similar way, and dividing out numerical common factors at each stage, Newton wrote down the following polynomials:

$$\begin{aligned}
&x^5 - 2x^4 - 10x^3 + 30xx + 63x - 120,\\
&5x^4 - 8x^3 - 30xx + 60x + 63,\\
&5x^3 - 6x^2 - 15x + 15,\\
&5xx - 4x - 5,\\
&5x - 2.
\end{aligned}$$

Now he looked for a value of x that will make all the above expressions positive. The value $x = 1$ is too small since in the fourth polynomial $5.1^2 - 4.1 - 5 = -4$, but $x = 2$ works (giving positive values 46, 79, 1, 7, 8, respectively). Therefore, Newton claimed, 2 is an upper bound for the positive roots. A similar procedure can be used to find a lower bound (in this case -3) for the negative roots.

Newton's procedure looks like repeated differentiation, and indeed is most easily understood as searching for a value of x beyond which every derivative is positive. It is more likely, however, that Newton arrived at his method by another route, namely, reducing all the roots by an amount sufficiently large that they all become negative. If all the roots are reduced by k, for example, that is, we make the transformation $y = x - k$, equation (24) becomes

$$\begin{aligned}
y^5 + 5ky^4 + 10k^2y^3 + 10k^3y^2 + 5k^4y + k^5&\\
-2(y^4 + 4ky^3 + 6k^2y^2 + 4k^3y + k^4)&\\
-10(y^3 + 3ky^2 + 3k^2y + k^3)&\\
+30(y^2 + 2ky + k^2)&\\
+63(y + k)&\\
-120 &= 0.
\end{aligned}$$

[71] Newton 1707, 252–255; 1720, 206–208.
[72] Newton 1707, 255–257; 1720, 208–210.

All the roots y of this equation are negative if every coefficient is positive, that is, if

$$\begin{aligned}&k^5 - 2k^4 - 10k^3 + 30k^2 + 63k - 120 > 0,\\&5k^4 - 8k^3 - 30k^2 + 60k + 63 > 0,\\&10k^3 - 12k^2 + 30k + 30 > 0,\\&10k^2 - 8k - 10 > 0,\\&5k - 2 > 0.\end{aligned}$$

Apart from some scalar multipliers these are precisely Newton's conditions. They are satisfied when $k = 2$, and therefore no root of (24) can be larger than 2.

Newton's method gives the outer limits for the set of all the roots, but not the intermediate limits for the individual roots. Although the *Arithmetica universalis* was published seventeen years after Rolle's *Traité d'algebre* of 1690, it had been written six or seven years before it, in 1683 or 1684, so that Newton could not then have known of Rolle's method of cascades for finding individual limits. His method does not give such precise information as Rolle's but nevertheless significantly restricts the range over which the roots must be sought.

Summary of Part I

The three chapters of Part I of this book have presented an overview of the main developments in the theory of equations from 1545 to the end of the seventeenth century. I have not attempted to survey every book or paper published. Many textbooks throughout this period, for example, offered basic instruction in writing and solving equations but they rarely went further than the standard rules for quadratic or occasionally cubic equations. My concern here has been, rather, to identify and explain ideas that in their time were new, and that were to be particularly significant later. Here it is perhaps useful to give a summary of those that were to prove most important.

1. Solution methods for cubic and quartic equations. These were set out by Cardano in the *Ars magna* in 1545, and it was surely these methods that Lagrange was referring to when he claimed in 1771 that there had been no advances since then. In the century and a half following the publication of the *Ars magna*, similar procedures for solving fifth or higher-degree equations had proved unaccountably elusive.

2. Results on the number and nature of the roots. In the first chapter of the *Ars magna* Cardano had already classified the number of positive or negative roots for each kind of cubic. In 1637 Descartes produced his 'rule of signs' for the number of positive roots of any equation, and in 1707 Newton published a 'rule of inequalities' for the number of 'impossible' roots, but both these rules remained unproved. As it turned out, neither was central to the later theory of equations but both were to give rise to a good deal of further work in the eighteenth century and beyond (see Chapter 4).

3. Roots as sums of radicals. Cardano had observed that solutions to quadratic equations consist in general of sums of rationals and square roots, while solutions to cubic equations consist of sums of rationals and pairs of cube roots. He made no explicit statement, however, about what one might expect to find inside those square or cube roots. A century later, Dulaurens solved some special fifth, seventh, and eleventh degree equations using sums of pairs of fifth, seventh, and eleventh roots respectively, but again without being explicit about what might appear inside the radical signs. Leibniz too appears to have investigated equations whose roots are sums of radicals, but ran into difficulty in the calculations. The idea that an equation of degree n might have roots expressible, at least in their outer layer, as sums of radicals of degree n was to become important later (see Chapter 5).

4. Transformation of equations. Cardano had taught the basic transformations $x \to x \pm k$ and $x \to k/x$ for either simplifying an equation or transforming it into a recognizably solvable case.

5. Removal of terms. The most common use of the above transformations was to remove the second highest term from a cubic or quartic. Dulaurens in 1667 pointed out that it was possible to remove *any* term, and indeed thought that it might even be possible to remove *all* intermediate terms, though he did not himself see how to do it. Tschirnhaus in 1683 published one possible way of proceeding, and Gregory in private tried another, but, as with the idea of roots as sums of radicals, the difficulty of the calculations quickly blocked any real progress (see Chapter 5).

6. Polynomials as products of factors. A paradigm shift in the understanding of equations came about between the sixteenth and seventeenth centuries through the discovery, first published in Harriot's writings (1631) and also illustrated by Descartes (1637), that polynomials could be construed as products of linear factors. This led rapidly to the almost complete abandonment of older ideas of equations as proportional relationships.

7. Composition of coefficients of the roots. In Harriot's systematic treatment of polynomials and their composition, it became clear how the coefficients of an equation were constructed from its roots, and under what conditions one or more terms would disappear.

8. Information about the roots from the coefficients. From the early seventeenth century it was known how the coefficients of an equation were constructed from the roots. The converse problem, of deriving individual roots from the coefficients by means of algebraic operations, remained the primary objective of equation solving. By the end of the seventeenth century no-one was any nearer than Cardano had been to achieving that for equations of degree higher than four. Much useful information, however, *could* be extracted from the coefficients: sums of powers of the roots, for instance, and also the limits or bounds within which the roots must lie (see Chapters 6 and 9).

9. Simultaneous solution of two equations. Hudde in 1658 had shown a method of solving two polynomials simultaneously, that is, of discovering common factors. Newton had given conditions under which two quadratics or cubics could have roots in common. These were only the most elementary cases of what later became known as the theory of 'elimination', which was to become a major area of research half a century later (see Chapter 7) .

The beginnings of almost every development in the eighteenth century can be traced back to one or other of the above discoveries made in the sixteenth or seventeenth centuries. The continuation of these ideas into the eighteenth century will be discussed in Part II. There the chronological approach followed in Part I will give way to chapters devoted instead to some of the separate strands and themes identified above.

Part II

From Newton to Lagrange: 1707 to 1771

Chapter 4
Discerning the nature of the roots

In Part II of this book we will follow chapter by chapter some of the individual themes that were identified and summarized at the end of Part I. The story from now on will therefore be shaped by interweaving threads, each of them traced from the beginning of the eighteenth century to some appropriate end point a few decades later. The material lends itself to this style of exposition because the theory of equations, like most good mathematical concepts, developed from a tangle of interconnected ideas that were only later seen to be part of a coherent whole.

Before exploring particular themes, however, we need first to re-orient ourselves, because the context within which mathematics was done and disseminated in the eighteenth century was already very different from that of the seventeenth. In the English, French, and German-speaking countries of western Europe, able and well-read mathematicians, though still relatively few in number, began to form professional communities. In continental Europe exchange of ideas was fostered by the Academies in Berlin (founded in 1700) and St Petersburg (founded in 1725), both modelled on the prototype at Paris. The Academies not only held discussions of mathematics within their meetings but also published mathematical papers in their respective journals, thus creating a permanent and public record of new advances and providing a forum for the exchange of ideas at more than merely local level. Where most seventeenth-century developments in the theory of equations had first appeared in books, those of the eighteenth century were more likely to be published in the *Mémoires* of the Academies of Paris or Berlin, the *Commentarii* or *Novi commentarii* of the Academy of St Petersburg, or the *Philosophical Transactions* of the Royal Society in London.

Such papers did not always draw a quick reaction: it could be months or years before anyone responded with new arguments or further developments. Part of the reason for this was that the eighteenth century saw mathematical advances in a multitude of directions, many of them on questions that were more easily answered or more immediately fruitful than the seemingly intractable problem of finding solutions to higher degree equations. Thus, at least during the first half of the eighteenth century, progress in the theory of equations came rather slowly and in somewhat piecemeal fashion. Euler in particular could be relied on to throw out clever ideas but he would then often abandon them as he turned to something else, leaving it for others to take up his work, sometimes not until many years later. It was this relatively slow development that makes it possible to trace individual themes over several decades until they finally came together in the 1770s.

This first chapter of Part II takes up one of the earliest problems identified in the theory of equations, already explored by Cardano in some detail in the first chapter of the *Ars magna*: without solving an equation what information can we discover about its roots? How many are there likely to be? How many will be real and how many imaginary?[1] And of those that are real, how many will be positive and how many negative? The answer to the first question, how many roots an equation might have, was already becoming clear in the sixteenth century. It had been known for centuries that a quadratic equation could have two roots and there was growing evidence that a cubic equation could have as many as three roots and a quartic as many as four. By the early seventeenth century it had become an accepted fact, based on evidence and intuition, that an equation has as many roots as its degree.

Questions about the nature of the roots, whether real or imaginary, positive or negative, were much harder to answer. The *Ars magna* already provided complete criteria for cubic equations, and later writers, Harriot in particular, offered partial results for quartics, but beyond that the problem became much more difficult. The first useful rule for gleaning information about equations of higher degree was Descartes' rule of signs for estimating the numbers of positive and negative roots (1637). The second was Newton's rule for estimating the number of imaginary roots (1707). Both of these rules, however, were to cause perplexity and discussion well into the eighteenth century. Descartes stated his rule of signs without proof, while Newton offered no general statement at all, only a few examples. In both cases their demonstrations by example left some important awkward cases undecided.

This chapter describes some of the eighteenth-century efforts to prove the rules of Descartes and Newton. Proofs or lack of them did not impinge to any great extent on other approaches to equation-solving, so the problem never evoked any concerted effort. Rather, it was something that almost any mathematician could turn his hand to, and attempts came and went depending on individual enthusiasms, resulting in scattered and isolated approaches. It is perhaps not surprising that Descartes' rule was taken up only by continental writers, while Newton's rule was pursued mainly by British mathematicians, though later also by Euler.

There is no happy ending to this chapter, but it outlines at least some of the work done on a problem that seemed as though it should be simple, but turned out to be obstinately difficult in practice.

Descartes' rule of signs, 1637 to 1740

There are two stories to untangle concerning Descartes' rule of signs, one mathematical, one historical. We will begin with the historical confusion. As we saw above (pages 46–47), Descartes wrote in his *Géométrie* of 1637 that

> one may have as many true roots as the number of times the signs $+$ and $-$ are found to change; and as many false roots as the number of times the two signs $+$ or the two signs $-$ are found to follow one another.

Until the late seventeenth century there was no question of attributing this rule to anyone but Descartes. Confusion was introduced, however, by John Wallis in his account of

[1] In this chapter we will follow eighteenth-century terminology and use the description 'imaginary' for roots that have both real and imaginary parts; today such roots would be described as 'complex'.

the work of Harriot in his *Treatise of algebra* (1685). There Wallis showed correctly how Harriot had been able to estimate the number of real roots of cubics and some quartics by comparing a given equation with canonical forms.[2] Wallis then claimed, in a statement far more general than Harriot's work actually allowed, that Harriot had been able to do this for *any* equation:

> And in this manner, in any Common equation proposed, by comparing it with a Canonick like Graduated, like Affected, and like Qualified (as to the respective Equality, Majority or Minority of its parts [coefficients] duly compared,) it will appear what number of real roots it hath, and how Affected.

Wallis later called this 'Harriot's Rule' though in truth it was not a rule at all, but rather a method of investigation that had worked for some particular lower degree equations.

Immediately following his discussion of Harriot's method, Wallis next presented the rule of signs, but without any mention of Descartes:[3]

> Now, (upon a survey of the several forms,) it will be found, that (the Equation being put all over to one side, and set in order;) as many times as in the order of Signs $+\ -$, you pass from $+$ to $-$, and contrariwise; so many are the Affirmative Roots: But as many times as $+$ follows $+$, or $-$ follows $-$; so many are the Negative Roots.

Wallis went on to point out that this rule holds only when all the roots are real. And, he claimed, to discover whether all the roots are real or not one needs what he now called 'Harriot's Rule':

> But how many of these be Real, and how many but Imaginary will depend upon that other condition of *Harriot*'s Rule; *viz.* that the compared Equations be *duly qualified*, as to the Equality, Majority, or Minority of their respective parts.

Only now did Wallis also name Descartes, but in such a way as to suggest that Descartes merely *agreed* with the rule of signs, rather than that he was its originator. At the same time Wallis could not resist also castigating him for not warning that all the roots must be real:

> As to the former of these [the rule for the number of positive roots], we have *Des Cartes* concurrence, (but without the caution interposed, which is a defect:] Of the latter [the rule for the number of real roots], (if I do not mis-remember) he is wholly silent.

[2] Wallis 1685, 157–158.
[3] Wallis 1685, 158.

To present the rule of signs within an account of Harriot's work and then claim that Descartes simply 'concurred' with it was at best misleading, at worst duplicitous. Unfortunately, however, the misapprehension that Harriot was the author of the rule of signs persisted. When Wallis's *Treatise of algebra* was reviewed in the *Acta eruditorum* in 1686 the anonymous reviewer wrote the following:[4]

> [Harriot] was the first to observe, by induction, as it seems, that there are as many negative roots (at least in an equation having its roots purely real or possible, a warning that Descartes in his other writings incorrectly omits) as there are changes of sign immediately following each other; and as many positive roots as agreements of the same.

Clearly the writer had paid little attention to the details of the matter. Not only did he state the rule of signs the wrong way round, but also attributed to Harriot something that is not to be found in any of his writings, manuscript or published. It is not difficult to see, however, how such misunderstanding arose from Wallis's text, especially if the reader was not completely fluent in English. It is very likely that the reviewer was Leibniz, who may also have conflated the written contents of the *Treatise of algebra* with snippets of conversation with Pell or Wallis on the subject of Harriot's algebra, half remembered from his visit to London some thirteen years previously. Unfortunately, his attribution of the rule of signs to Harriot became the accepted story, repeated by many eighteenth-century writers.[5]

The second confusion around Descartes' rule is a mathematical one, concerning the conditions under which the rule of signs holds. Descartes never claimed that the rule would predict *exactly* the number of positive roots but only the number that there *might* be: 'one may have as many true roots ...' ('*il y en peut avoir autant de vrayes ...*'). On the other hand he offered as an example the equation $+x^4-4x^3-19xx+106x-120=0$, with roots 2, 3, 4, -5. Here he claimed that one knows that there *are* three true roots ('*on connoist qu'il y a trois vrayes racines*'). In this case the rule gives the number of positive (and negative) roots precisely, because all the roots are real. This last is a necessary condition if the rule is to give the actual rather than possible number of positive roots, but Descartes nowhere stated it.

As early as 1659 Frans van Schooten pointed out that there may be fewer positive or negative roots than the rule suggests.[6] That did not prevent others from commenting

[4]*Observavit primus, ex inductione, ut videtur, tot esse radices privativas (in aequatione scilicet meras radices reales seu possibiles habente, quam cautionem Cartesius caeteris descriptis non recte omisit) quot sunt mutationes signorum immediate sibi succedentium; tot positivas, quot eorundem consensus;* Anonymous 1686, 285.

[5]Von Wolff 1739, 202–203; Saunderson 1740, II, 683; Kästner 1761; for discussion of the attribution see de Gua 1741a, 74–76.

[6]*ut qualibet Aequatione non tot radices haberentur, quot incognita quantitas habet dimensiones; neque tot verae, quot in ea reperientur variationes signorum + & −; aut tot falsae, quot vicibus deprehenduntur duo signa + vel duo signa −, quae in se invicem sequantur.* [so that in any equation there may not be so many roots as the unknown quantity has dimensions, nor may there be so many true roots as there are variations of sign + and −, nor as many false as in turn there are found two + signs or two − that may follow each other.] Descartes 1659, I, 285–286.

on Descartes' omission. In 1684, Michel Rolle complained anew that Descartes' rule was not general, leading the Paris Academy to investigate the matter and to report that van Schooten had already made the same observation.[7]

It was easier to engage in arguments about the authorship or veracity of the rule of signs than to prove it, and there was no published proof until the 1740s. Well before that there had been at least some progress on verifying Newton's rule.

Newton's rule for imaginary roots, 1707 to 1730

Descartes' rule enables one to make at least a little progress with discovering how many roots are positive or negative. Determining how many roots are real and how many imaginary, however, is much more difficult. As we saw earlier (page 73), Newton had presented a procedure for doing so in the *Arithmetica universalis*, but without any explanation of why it should work. There we saw Newton demonstrating the method on an equation of degree 5. Here, as a reminder of his algorithm, is another of his examples, this time of degree 3. Consider the equation

$$x^3 + pxx + 3ppx - q = 0.$$

Take the fractions $\frac{1}{3}$, $\frac{2}{2}$, $\frac{3}{1}$, divide each by the one that follows, and write the results above the inner terms of the equation, thus

$$\begin{array}{ccccc} & \frac{1}{3} & \frac{1}{3} & & \\ x^3 & +pxx & +3ppx & -q & = 0. \\ + & - & + & + & \end{array}$$

The sign below a term is then + or − according to whether the square of the term, multiplied by its overhead fraction, is greater or less than the product of the terms on either side of it. In this case $\frac{1}{3}ppx^4 < 3ppx^4$ but $\frac{1}{3}9p^4xx > -qpxx$. A + sign is placed under each of the end terms. Newton claimed that each change of sign from + to − or − to + indicates the existence of an 'impossible' root. In this case, therefore, one would expect two such roots corresponding to the changes + − and − +.

The first published attempt to justify this rule came from Colin Maclaurin who, with Newton's support, had been appointed to the chair of mathematics at Edinburgh in 1725. Maclaurin began working on a proof of Newton's rule that same year and in the spring of 1726 sent his preliminary findings to Martin Folkes, vice-president of the Royal Society, with whom he was in regular correspondence. Apparently without consulting Maclaurin, Folkes in turn passed the letter to the secretary, James Jurin, and it was published in the *Philosophical Transactions* for May 1726 under the title 'A letter from Mr Colin Maclaurin [...] concerning aequations with impossible roots'.[8]

[7]See *Journal de l'Académie des Sciences* (1684), 20; Prestet 1694, II, 362–366; de Gua 1741a, 76–77.

[8]Maclaurin (1726–27). Maclaurin's account of the publication of this paper is to be found in Mills 1982, 224.

Maclaurin's work was based on a simple algebraic inequality; if $a, b, c, \ldots$ are m real quantities, then

$$(m-1)(a^2+b^2+c^2+\cdots) > 2ab+2ac+2bc+\cdots. \qquad (1)$$

Maclaurin proved this by summing of the squares of the differences of the m quantities. The argument is given here in Maclaurin's notation, in which bracketing of terms is represented by overlines, for example, $\overline{a-b}$. Clearly the sum of squares is positive, that is,

$$\overline{a-b}^2+\overline{a-c}^2+\overline{b-c}^2+\cdots > 0.$$

On the left, each square $a^2, b^2, \ldots$ occurs $m-1$ times but each product of the form $-2ab$ just once, and so (1) follows. By taking $a, b, c, \ldots$ to be real roots of equations and applying appropriate versions of (1), Maclaurin was able to confirm Newton's rules for quadratic, cubic, and quartic equations, together with a partial result for equations of higher degree. The published paper ends abruptly, however, with the words 'To be continued'.

Two years later a much longer and more detailed paper on the same subject was published in the *Philosophical Transactions* by George Campbell, also from Edinburgh.[9] In 1725 Campbell had been a rival to Maclaurin for the chair of mathematics, and the publication of their respective papers gave rise to a brief but unpleasant controversy. The story can be pieced together from the surviving correspondence between Maclaurin and his friend (and fellow Scot) James Stirling, then in London and active within the Royal Society. According to Stirling's later account, Campbell had sent his paper to the mathematician John Machin very soon after Maclaurin's paper was printed in the spring of 1726, but Machin being busy with other matters was slow to attend to it. Maclaurin, on the other hand, recalled that he had spent a day with Machin in September 1727 and that Machin had made no mention of any such paper. Campbell himself claimed that he had sent it in the autumn of 1727. Maclaurin argued, however, based on his memory of a conversation with Campbell in August 1728, that Campbell had not in fact sent it until June 1728 at the earliest.[10]

It was in the course of that same conversation in August 1728 that Maclaurin discovered that Campbell had found a method of demonstrating Newton's rule by considering the 'limits' or bounds between which the roots must lie (see below).[11] By then Maclaurin too had discovered such a method and was therefore disconcerted to discover from Stirling later that autumn that Campbell's paper was to be published in the *Philosophical Transactions* for October. By December, he had still not seen it, but wrote to Stirling in a state of mild concern, admitting that he had taken far too long to complete his own work, but wishing that Campbell's paper might have been held back in view of the fact that his own paper was so obviously unfinished. He also took the precaution of outlining for Stirling his own argument based on limits.[12]

[9]Campbell 1727–28.
[10]Mills 1982, 183, 188, 240.
[11]Mills 1982, 185, 240.
[12]Mills 1982, 181.

Stirling explained in his reply that Campbell's paper had been published as a result of intense pressure from one of Campbell's supporters, Sir Alexander Cuming, and gave Maclaurin a very brief summary of its contents.[13]

It was not until the beginning of February 1729 that Maclaurin saw the article for himself. Recognizing that some of Campbell's results overlapped with his own findings and hoping to avert a quarrel Maclaurin immediately drafted a letter to Campbell:[14]

> I send you this Theorem (meaning my Sixth Proposition) because you will see it was impossible for me to find it out since Eleven last Night, when I first saw your Paper. I have also drawn from it, many other Consequences, besides what you have in your Paper; all which, when you see them, will more fully satisfy you, that these Theorems lay in the Way I had taken, so I actually had them. My only Design in sending you this Note, is to prevent any Dispute or Misunderstanding about this Affair, as much as I can.

Unfortunately, this letter was never sent. On the advice of a 'Professor in our University', and conscious of his earlier rivalry with Campbell, Maclaurin abandoned it.[15] Instead, a week later, he wrote a lengthy letter to Stirling expressing his concern that Campbell had pre-empted him by taking up the ideas he himself had set out:[16]

> I cannot therfor [sic] but be a little concerned that after I had given the principles of my Method and carried it some length and had it marked that my paper was to be continued another pursing the very same thought should be published at the intervall;

In fact it was not true that Campbell had taken his lead only from Maclaurin's first paper for, as Maclaurin had deduced from their conversation in August 1728, Campbell had introduced a different idea into the discussion. Suppose the proposed equation is (using Campbell's notation)

$$x^n - Bx^{n-1} + Cx^{n-2} - \cdots \pm cx^2 \mp bx \pm A = 0.$$

From this we can obtain a second equation,

$$nx^{n-1} - \overline{n-1}Bx^{n-2} + \overline{n-2}Cx^{n-3} - \cdots \pm 2cx \mp b = 0.$$

Campbell explained that the second equation is formed from the first by multiplying each term by its exponent and dividing by x, that is, he regarded this as a purely algebraic process, not an application of calculus. Suppose that all the roots of the first equation are real. Then, claimed Campbell, so are all the roots of second (though the converse need not be true). For justification Campbell referred his readers to demonstrations by 'Algebraical writers, particularly by Mr. *Reyneau* in his *Analyse*

[13] Mills 1982, 183–184.
[14] Mills 1982, 230.
[15] Mills 1982, 230, 423.
[16] Mills 1982, 185–188.

demontré [sic]'.[17] As it happened, Maclaurin had made the same statement in his *Treatise of algebra*, where he referred to the roots of the second equation as 'limits', that is, boundaries, between the roots of the first equation.[18] The *Treatise of algebra*, though not published until 1748, was written for the most part during 1726. Consisting as it did essentially of Maclaurin's lecture notes it was, as Maclaurin later pointed out, 'very publick in this Place [Edinburgh]'. Even before the argument with Campbell, Maclaurin was concerned that its contents would be taken up by others because 'my dictates go through every body's hands here'.[19] Whether Campbell had picked up the idea of using 'limits' from Maclaurin's notes or, as he claimed, from Reyneau's *Analyse démontrée* is impossible to say.

Continuing Campbell's procedure as above, one eventually arrives at a quadratic equation:

$$n\overline{\frac{n-1}{2}}x^2 - \overline{n-1}Bx + C = 0. \tag{2}$$

This too must have all its roots real, so it must be the case that

$$\overline{\frac{n-1}{2n}}B^2 > 1 \times C. \tag{3}$$

Campbell justified this from the usual formula for the roots of (2); whereas Maclaurin had proved it using inequality (1).

Now if the original equation has n real roots then so does the equation obtained from it by substituting $y = 1/x$, namely,

$$Ay^n - by^{n-1} + cy^{n-2} - \cdots \pm Cy^2 \mp By \pm 1 = 0.$$

A further application of Campbell's argument leads again to condition (3) but now between c, b, and A:

$$\overline{\frac{n-1}{2n}}b^2 > c \times A. \tag{3'}$$

Having established conditions (3) and (3′) for the three coefficients at either end of the equation, Campbell went on to investigate the relationship that must hold between any three consecutive coefficients L, M, N. This he did by writing down the sequence of 'ascending' equations that precedes (2), the first of which was

$$n \times \overline{\frac{n-1}{2}} \times \overline{\frac{n-2}{3}}x^3 - \overline{n-1} \times \overline{\frac{n-2}{2}}Bx^2 + \overline{n-2}Cx - D = 0.$$

Continuing in the same way, he arrived at

$$\begin{aligned} &n \times \overline{\frac{n-1}{2}} \times \overline{\frac{n-2}{3}} \cdots \times \overline{\frac{n-m}{m+1}}x^{m+1} + \cdots \\ &\pm \overline{n-m+1} \times \overline{\frac{n-m}{2}}Lx^2 \mp \overline{n-m}Mx \pm N = 0. \end{aligned} \tag{4}$$

[17] Reyneau 1708.

[18] *If any Roots of the Equation of the Limits are impossible, then must there be some Roots of the proposed Equation impossible.* Maclaurin 1748, 182.

[19] Mills 1982, 182, 240.

Now applying condition (3′) to the last three terms of (4) gave him

$$\frac{m \times \overline{n-m}}{\overline{m+1} \times \overline{n-m+1}} M^2 > L \times N. \tag{5}$$

Campbell called this Proposition I, the first of the two he put forward in his paper.

He then explained that Newton's rule begins from a sequence of fractions

$$\frac{n}{1}, \quad \frac{n-1}{2}, \quad \frac{n-2}{3}, \quad \frac{n-3}{4}, \ldots,$$

each of which is to be divided by the preceding one to give the sequence

$$\frac{n-1}{2n}, \quad \frac{2 \times \overline{n-2}}{3 \times \overline{n-1}}, \quad \frac{3 \times \overline{n-3}}{4 \times \overline{n-2}}, \ldots.$$

These are then placed over successive terms of the equation. Thus Campbell's condition (5) is precisely Newton's condition for writing + under the term with coefficient M. The reversal of the inequality signifies the existence of a pair of imaginary roots; Newton's rule requires that in this case we insert −.

Campbell's Proposition II refined and strengthened Proposition I. Suppose that $\ldots I$, K, L, M, N, O, P, ... are successive coefficients of an equation in which M is the coefficient of x^m. Then, he claimed, if all the roots are real it will be the case that

$$M^2 \times \frac{1}{2} \times \overline{1 - \frac{1}{n \times \frac{n-1}{2} \times \frac{n-2}{3} \times \cdots \times \frac{n-m+1}{m}}} > L \times N - K \times O + I \times P - \cdots. \tag{6}$$

To derive this, Campbell, like Maclaurin, used an argument based on sums of differences of squares. His condition (6) identifies the existence of imaginary roots more effectively than (5) but is more laborious to apply. Campbell demonstrated the relative strengths of (5) and (6) by applying each in turn to the equation

$$x^7 - 5x^6 + 15x^5 - 23x^4 + 18x^3 + 10x^2 - 28x + 24 = 0.$$

Recall that Newton (and Campbell) were interested not just in the sign to be placed under each term but in the *sequence* of signs, and in particular in any changes from + to − or vice versa. Condition (5) (Newton's rule) predicts only two imaginary roots for the above equation, whereas condition (6) (Campbell's rule) correctly predicts six (the roots are -1, $1 \pm \sqrt{-1}$, $1 \pm \sqrt{-2}$, $1 \pm \sqrt{-3}$).

It was Campbell's Proposition II that so alarmed Maclaurin when he saw it in print in February 1729. He demonstrated immediately to Stirling that he too could produce the same theorem and even better ones, and argued that he could not possibly have done so in the short time since reading Campbell's paper:[20]

[20]Mills 1982, 187.

> I believe you will easily allou I could not have invented these theorems since tuesday last especially when at present by teaching six hours daily I have little relish left for such investigations. […] I am afraid these things are not worthy your attention. Only as these things once cost me some pains I cannot but with some regret see myself prevented.

Stirling suggested that the best way for Maclaurin to proceed was to publish his own findings as rapidly as possible. Maclaurin began revising his work as soon as he had leisure to do so and on 19 April sent the completion of his paper to Martin Folkes. On 1 May he wrote to Stirling again, now in a rather calmer state of mind.[21]

> I find he [Campbell] has prevented me in one Proposition only; which I have shoued without naming or citing him or his paper to be the least Valuable. […] I am sorry to find you so uneasy about what has hapenned in our last letter. It is over with me. When I found one of my Propositions in his paper I was at first a little in pain; but when I found it was only one of a great many of mine that he had hit upon; and reflected that the generality of my Theorems would satisfy any judicious reader; I became less concerned. All I nou desyre is to have my paper or at least the first part of it published as soon as possible.

Maclaurin's paper was indeed published rapidly, in the *Philosophical Transactions* for March and April 1729, under the title 'A second letter from Mr Colin Maclaurin […] concerning the roots of equations'.[22] Taking up from where he had left off in 1726, Maclaurin began his second paper at Proposition VI (the proposition he had written out for Campbell back in February but did not send). His Proposition VII is identical to Campbell's Proposition II but Maclaurin's version is more simply written because he set $l = n \times \frac{n-1}{2} \times \frac{n-2}{3} \times \frac{n-3}{4} \times \frac{n-4}{5}$. Maclaurin asserted that if a polynomial of degree n has successive coefficients $1, -A, +B, -C, +D, -E, F, \ldots$, for example, it will have a pair of imaginary roots if

$$\frac{l-1}{2l} E^2 < DF - CG + BH - AI + K,$$

which is simply the converse of what Campbell had claimed slightly more generally in his Proposition II (see (6)). Maclaurin's proof of Newton's rule came in Proposition IX, which is therefore equivalent to Campbell's Proposition I (see (5)).

Maclaurin did not stop there but proceeded to yet more complicated rules in Propositions XI and XII which, he claimed, were better than those in Propositions VII and IX and could sometimes discover imaginary roots when those could not. He also discussed the possibility that there might be more than a single pair of imaginary roots, as, for example, in the equation

$$x^4 - 4ax^3 + 6a^2x^2 - 4ab^2x + b^4 = 0.$$

[21] Mills 1982, 215–216.

[22] A shorter presentation of some of the results from this paper can also be found in Maclaurin 1748, 274–285.

Maclaurin's Proposition IX (Newton's rule) applied to this equation indicates the existence of four imaginary roots when $a > b$ (actually when $a^2 > b^2$ but Maclaurin took a and b to be positive). Proposition VII indicates four imaginary roots when $a > b$ or $b^2 > 15a^2$. Proposition XI indicates four imaginary roots when $a > b$ or $b^2 > 9a^2$. Thus the conditions in Propositions IX, VII, XI are increasingly refined.[23]

Finally, at the end of his paper, Maclaurin returned to simpler rules based on the following theorem.

THEOREM III. *In general the Roots of the Equation*

$$x^n - Ax^{n-1} + Bx^{n-2} - Cx^{n-3}\,\&c. = 0$$

are the Limits *of the Roots of the Equation*

$$nx^{n-1} - \overline{n-1}Ax^{n-2} + \overline{n-2}Bx^{n-3}\,\&c. = 0,$$

or of any Equation that is deduced from it by multiplying its Terms by any Arithmetical Progression $l \pm d, l \pm 2d, l \pm 3d$ &c. and conversely the Roots of this new Equation will be the Limits *of the Roots of the proposed Equation*

$$x^n - Ax^{n-1} + Bx^{n-2} - Cx^{n-3}\,\&c. = 0.$$

This theorem had already been proved by Rolle (see pages 69–70). Possibly Maclaurin, like Campbell, had studied Reyneau's *Analyse demontrée* where part of it is quoted.[24] Maclaurin, however, now proved it again for himself using the lemmas he had established earlier.

It follows from Theorem III that if the given equation is

$$x^n - Ax^{n-1} + Bx^{n-2} - Cx^{n-3} + \cdots = 0, \tag{7}$$

then the existence of imaginary roots for any derived equation, for example,

$$nx^{n-1} - \overline{n-1}Ax^{n-2} + \overline{n-2}Bx^{n-3} - \cdots = 0 \tag{8}$$

[23] The discussion in Mills 1982, 201, n 221, is confused on this subject and not always correct. First, when Maclaurin wrote in relation to Proposition VII: 'when a is greater than b and also when b^2 is greater than $15a^2$' he was offering alternatives, not claiming that both conditions could hold at once. Second, where there is more than one pair of imaginary roots one needs to examine not only the individual signs under each term but the *succession* of signs. Third, in considering such a succession, one sees that Proposition VII indicates the existence of a second pair of imaginary roots when $15a^4 > 16a^2b^2 - b^4$ or $(15a^2 - b^2)(a^2 - b^2) > 0$, that is, when $a^2 > b^2$ or $b^2 > 15a^2$; Proposition XI indicates a second pair of imaginary roots when $9a^4 > 10a^2b^2 - b^4$ or $(9a^2 - b^2)(a^2 - b^2) > 0$, that is, when $a^2 > b^2$ or $b^2 > 9a^2$, exactly as Maclaurin asserted.

[24] *si on multiplie les termes d'une équation quelconque, dont tout les racines sont réelles, positives & inégales, chacun par le nombre qui est l'exposant de l'inconnue de ce terme, & le dernier terme par zero, les racines de l'équation qui vient de cette multiplication, sont les limites des racines de l'équation proposée.* [if one multiplies each of the terms of any equation in which all the roots are real, positive, and distinct by the number which is the exponent of the unknown in that term, and the last term by zero, the roots of the equation that comes from this multiplication are the limits of the roots of the original equation.] Reyneau 1708, I, 290.

or

$$Ax^{n-1} - 2Bx^{n-2} + 3Cx^{n-3} + \cdots = 0, \tag{9}$$

implies the existence of imaginary roots for (7). Since both (8) and (9) are of lower degree than (7) they may be easier to investigate than (7) itself.

Now suppose that $Dx^{n-r+1} - Ex^{n-r} + Fx^{n-r-1}$ are three consecutive terms of (7). By continuing the procedure that took us from (7) to (8) above, we can eliminate all terms to the right of the term containing F (as Campbell had done). We can then multiply the resulting equation by 0, 1, 2, 3, ... as often as necessary to eliminate terms to the left of the term containing D (a refinement of Campbell's method). In this way Maclaurin arrived at the quadratic equation

$$\overline{n-r+1} \times \overline{n-r} \times 2Dx^2 - 4\overline{n-r} \times rEx + 2\overline{r+1} \times rF = 0,$$

and the condition for this to have imaginary roots is

$$\frac{\overline{n-r} \times r}{\overline{n-r+1} \times \overline{r+1}} E^2 < DF.$$

This was the condition that Maclaurin had already given at Proposition IX and it was also Campbell's Proposition I (see (5); Maclaurin's $n - r$ here is just Campbell's m there). Maclaurin's second method of deriving it was very similar to Campbell's, but whether influenced by it or not is hard to say. It is likely that both had come close to Newton's original derivation.

That might have been the end of the story except that Campbell was rather less magnanimous towards Maclaurin than Maclaurin had been towards him. In October of that year Campbell printed a pamphlet entitled *Remarks on a paper published by Mr. MacLaurin, in the Philosophical Transactions for the month of May, 1729* (his date was wrong) in which he complained bitterly that Maclaurin had accused him of plagiarism, and further that there were errors in Maclaurin's paper.[25] Maclaurin wrote a lengthy reply, which was also printed: *A defence of the letter published in the Philosophical Transactions for March and April 1729.*[26] Maclaurin argued that he had not mentioned Campbell's name in his published paper nor accused him of plagiarism in any conversation or letter, though he had confessed he had complained to an acquaintance that 'there always arose great Inconveniencies from a Person's interfering with any One in what he has begun and carried some Length, when he promises the Sequel ...'. But, as Maclaurin reasonably pointed out, there was a difference between 'pursuing a method begun by another' and actual plagiarism. He also observed that 'the subject of our papers was abstruse' and that it was therefore perhaps difficult for Campbell's casual acquaintances to understand the precise nature of the similarities or differences between them. Finally, he wrote at length on the perceived errors in his paper.

Clearly here were two men with an interest in the same question who were both able to make significant progress, but publishing delays and a past history of rivalry

[25]Campbell 1729.

[26]Maclaurin 1730.

led almost inevitably to a priority dispute. With the benefits of hindsight one may summarize the position as follows. The key method in Maclaurin's first paper was his use of inequalities derived from sums of squares. Campbell also made use of such inequalities, but only in the second part of his paper; the first half was based instead on the algebraic theory of 'limits'. Campbell claimed that he had learned this from Reyneau, but Maclaurin had also already written on it in his *Treatise of algebra* in 1726. Thus both the key ideas, of algebraic inequalities and limits, had been identified by Maclaurin before Campbell worked with them. Nevertheless, Campbell saw their potential and made good use of them to deduce Newton's rule and to construct a further rule of his own, well before Maclaurin got round to writing up his full findings. We can now afford to be more generous than they could be and say that both deserve credit for their confirmations and extensions of a rule first published some twenty years earlier.

Descartes' rule of signs, 1741 onwards

The results discovered by Maclaurin and Campbell turned out also to be of some importance in creating a proof of Descartes' rule of signs. The first person to publish such a proof was Jean Paul de Gua de Malves, of whom little is known except that he appears to have been one of the early supporters of the creation of the *Encylopédie*. In 1741 he offered not just one but two proofs of Descartes' rule, in a paper published in the *Mémoires* of the Paris Academy.[27]

De Gua gave the name 'variation' (*variation*) to a succession $+-$ or $-+$ and 'permanence' (*permanence*) to a succession $++$ or $--$. In his first proof he considered any three consecutive terms of a polynomial of degree n, namely, $\pm Fx^{n-m} \pm Gx^{n-m-1} \pm Hx^{n-m-2}$, where the letters F, G, H are assumed to represent positive quantities. First de Gua re-proved result (5), which had already been proved by Maclaurin and Campbell in the late 1720s. He admitted that his result was the same as theirs, and that indeed his demonstration followed similar principles, but he argued that he wanted to set out a proof for his own purposes. De Gua's argument was in fact rather more straightforward than those of Campbell and Maclaurin, consisting simply of repeated multiplication by arithmetic progressions to eliminate all but three consecutive terms of the original equation. The result that de Gua arrived at, equivalent to (5) above, was

$$\frac{\overline{m+1}.\overline{n-m-1}}{\overline{m+2}.\overline{n-m}}.G^2 > FH.$$

Now, since $m+1 < m+2$ and $n-m-1 < n-m$ de Gua could deduce *a fortiori* that

$$G^2 > FH.$$

A corollary to this is that if p is a positive number then $pF > G$ implies that $pFG > G^2 > FH$ so that $pG > H$.

[27] De Gua 1741a.

Now de Gua examined the result of multiplying a polynomial by $(x+p)$, that is, of introducing a new negative root. Suppose the first variation in the original polynomial is $+Fx^{n-m} - Gx^{n-m-1}$. Multiplication of these two terms by $(x+p)$ gives

$$+Fx^{n-m+1} + (pF - G)x^{n-m} - pGx^{n-m-1}.$$

If $pF < G$ there is still a variation from the first to the second of these three terms. If $pF > G$ the variation changes to a permanence, but there will now be a variation from the second to third term regardless of whether the term Hx^{n-m-2} is positive or negative (because of the fact that $pF > G$ makes $pG > H$). Thus de Gua could argue that multiplying by a factor $(x+p)$, that is, introducing a new negative root, preserves the number of variations but increases the number of permanences by 1. Likewise, multiplying it by $(x-p)$, that is, introducing a new positive root, preserves the number of permanences but increases the number of variations by 1.

His second proof was rather different. Here he argued first that it is always possible to destroy one variation in an equation by multiplying term by term by an arithmetic progression containing the sequence ..., 1, 0, −1, ... (where 0 must multiply one of the terms contributing to the variation). In the second part of the argument he claimed that such multiplication creates a new equation with one fewer positive root than the original. Continuing far enough one reaches an equation with all its terms positive and therefore no positive roots. Thus the original equation can have had no more positive roots than variations (and, by an extension of the argument, no more negative roots than permanences).

For both proofs de Gua considered zero coefficients separately. The fact that such coefficients could be considered to be 'infinitely small positive' or 'infinitely small negative' led him to regard them as either positive or negative. Where such ambiguity leads to contradictions concerning the number of positive or negative roots, he argued, it can be taken as a sign of the existence of imaginary roots instead.

Another proof of Descartes' rule appeared in the *Mémoires* of the Berlin Academy in 1756, this time by Johann Andreas von Segner, professor of mathematics at Halle. Like de Gua in his first proof, von Segner investigated the effect of multiplying a given polynomial by a new factor of the form $(x+p)$ or $(x-p)$. As an example, he multiplied $+x^5 + 3x^4 - 5x^3 - 4x^2 + 12x - 13$ by $x+2$, multiplying first by $+x$ and then by $+2$ to obtain the two summands A and B,

$$\begin{array}{cccccccccccc}
\rightarrow & +x^6 & +3x^5 & & -5x^4 & -4x^3 & & +12x^2 & & -13x & & A \\
 & & & \searrow & & & \nearrow & & \searrow & & & \\
 & & +2x^5 & & +6x^4 & -10x^3 & & -8x^2 & & +24x & -26 & B.
\end{array}$$

The sign pattern in the final sum can be written down by following the arrows to obtain

$$+ \quad + \quad + \quad - \quad + \quad + \quad -.$$

❀ 296 ❀

ſoit univerſel. Soit premièrement une équation quelconque à multiplier par une équation ſimple, dont les ſignes ſoyent + +, & qu'on produiſe par cette multiplication des ſeries de ſignes, telles que celles que nous avons nommées au commencement A, B.

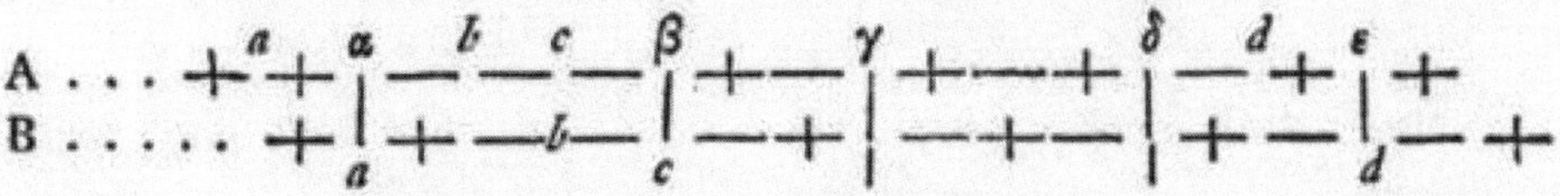

Toutes les ſucceſſions des ſignes ſemblables + + ou — — étant donc marquées par les lettres a, b, c, d, les mêmes dans l'une & dans l'autre ſerie, il eſt clair qu'on obtient indifféremment leur nombre, en comptant ces lettres, ſoit qu'elles ſe rencontrent dans la ſerie A, ou dans la ſerie B, pourvu ſeulement qu'on n'en prenne aucune deux fois. A' preſent, ſi en poſant les ſignes de l'équation produite, il faut paſſer en α de A en B, il faudra en β retourner de B en A; puis de nouveau en γ deſcendre de A en B, & ainſi de ſuite, en plaçant alternativement les ſignes des ſucceſſions ſemblables dans l'équation produite, laquelle tire ſes ſignes depuis le commencement jusqu'en α de la ſerie A, de α en β de la ſerie B, de β en γ, de nouveau de A, & toujours de même, tant qu'à la fin en ϵ on les prenne de B. En les recueillant de cette maniere, la premiere ſucceſſion eſt a; la ſeconde arrive néceſſairement auprès de α en deſcendant de A en B; la troiſième eſt b de la ſerie B; la quatrième ſeroit c de la même ſerie; mais celle-ci ſe détruit quelquefois en montant à la ſerie A, d'où vient que jusqu'ici il n'y a que trois ſucceſſions; mais la quatrième arrive auprès de γ en deſcendant, la cinquième auprès de δ en montant, la ſixième eſt en d, & la ſeptième a lieu auprès de ϵ en deſcendant. Cet exemple fournit donc le nombre des ſucceſſions ſemblables, augmenté de trois de ces ſucceſſions.

En général il paroit qu'à chaque deſcente de A en B le nombre de ces ſucceſſions augmente néceſſairement d'une; & qu'à chaque montée

A proof of Descartes's rule, from von Segner (1756).

Von Segner now noted that a movement from A to B or *vice versa* occurs only in the following sign patterns (the labels a, b, c, d are his):

$$\begin{array}{cccc} a\ + & & -\ c \\ & \times & \\ d\ + & & +\ b \end{array} \qquad \text{or} \qquad \begin{array}{cccc} a\ + & & -\ c \\ & \times & \\ d\ - & & +\ b, \end{array}$$

to which he should also have added

$$\begin{array}{cccc} a\ - & & +\ c \\ & \times & \\ d\ - & & -\ b \end{array} \qquad \text{or} \qquad \begin{array}{cccc} a\ - & & +\ c \\ & \times & \\ d\ + & & -\ b, \end{array}$$

which he did not state explicitly but used later. Von Segner gave no argument to justify his claims. We know, since all the multipliers are positive, that the sign at a must be the same as the sign at b. We can also see that an ascent or descent will occur only when the signs at positions c and b are different. The four patterns given above are therefore the only ones in which movement *can* take place, but it is not at all clear whether it *must* take place, and von Segner did not address this point. He was correct about the consequences, however: that a descent from A to B always introduces a repeated sign, while an ascent from B to A could either introduce or destroy a repetition. Since we always begin in A but end in B there are more descents than ascents, that is, multiplication by a factor of the form $x + p$ must add at least one new repetition. Von Segner used a similar argument to show that multiplication by a factor of the form $x - p$ must add at least one new change of sign. The rule of signs follows from this, giving precisely the number of positive and negative roots if all the roots are real.

Imaginary roots: examination of curves, 1717 to 1755

The rules given by Newton, Maclaurin, and Campbell for discerning imaginary roots were purely algebraic, based on the calculation of sums and products of the coefficients of the given equation. A different though parallel approach arose from examining graphs of polynomials, their 'serpentine curves'. The real roots of a polynomial equation correspond to the points where such a curve crosses the x-axis. Further, if all the roots are real and distinct, they will be separated from each other by positive local maxima or negative local minima. Where there are imaginary roots, however, this pattern breaks down. A local minimum that is positive rather than negative, for instance, indicates the existence of imaginary roots (as, for example, in the graph of $y = x^2 + 1$).

James Stirling made just such observations, in his *Lineae tertii ordinis Neutonianae* (1717), his commentary on Newton's classification of cubics.[28] Stirling examined all

[28]Stirling 1717, 59–68.

possible configurations of crossing points and turning points for curves corresponding to polynomials of degree two, three, and four, and discovered conditions under which each would have imaginary roots. He concluded that a quadratic equation of the form $x^2 + Bx + C = 0$ will have two imaginary roots if $B^2 - 4C < 0$, and that a cubic $x^3 + Bx^2 + Cx + D = 0$ will have two imaginary roots if $B^2 - 3C < 0$. For quartic equations, however, he was not able to come up with any single rule.

Some twenty years later de Gua referred several times to Stirling's work in his own paper on counting imaginary roots.[29] De Gua followed a similar approach to Stirling but now also brought the methods of calculus into play. If a polynomial $p(x)$ has a positive local minimum, then at that point $p(x) > 0$ and at the same time $p''(x) > 0$, and therefore $p(x)p''(x) > 0$. Similarly at a negative local maximum, $p(x) < 0$ and $p''(x) < 0$, and again $p(x)p''(x) > 0$. De Gua wrote this condition as $yddy > 0$ where $y = x^n + Bx^{n-1} + Cx^{n-2} + \cdots$. He claimed correctly that a polynomial has a pair of imaginary roots whenever the first derived equation has a real root *and* for that value of the root the condition $yddy > 0$ holds.[30]

The problem with this method is that it requires one to solve the derived equation, which is only one degree lower than the original. De Gua offered the following example.[31] Suppose we have an equation of degree 48 and we solve its first derived equation, of degree 47, to find 24 imaginary roots, 6 real positive roots and 17 real negative roots. Further, suppose that for one of the real positive roots and three of the real negative roots the condition $yddy > 0$ holds. One may conclude that the original equation has $24 + 8 = 32$ imaginary roots. By examining the possible positions of the remaining stationary points, where $yddy < 0$ (which occur at 5 positive values and 14 negative values of x), one may conclude that the original equation has at least $5 - 1 = 4$ real positive roots and at least $14 - 3 = 11$ real negative roots. This accounts for 47 roots, and the sign of the 48th and final root can be discovered by examining the product of all the roots. Such an argument, though valid, is, of course, completely impractical.

Both the geometric approach, through the study of curves, and the analytic approach, using differential calculus, give rise to the problem of what an imaginary root actually is. Both approaches demonstrate the existence of imaginary roots only as the absence of real roots. In 1746, Euler in his 'Recherches sur les racines imaginaires des equations' defined an imaginary quantity as one that is neither larger than zero, nor smaller than zero, nor equal to zero.[32] Several pages later Euler asserted that it was likely that every imaginary root was reducible to the form $M + N\sqrt{-1}$, and spent the next part of the paper proving that algebraic operations (addition, subtraction, multiplication, division, raising powers, taking roots) on numbers of the form $M + N\sqrt{-1}$ always lead to other number of the same kind, concluding that 'no operation can take us away from this form.[33] In the final part of the paper Euler showed that what he called transcendental

[29] De Gua 1741b.

[30] De Gua described a stationary point defined by $yddy > 0$ as a *minimum* and one defined by $yddy < 0$ as a *maximum*. The mismatch between this and modern terminology is so confusing that I have avoided de Gua's descriptions, and instead for each turning point have given the relevant inequality, which is unambiguous.

[31] De Gua 1741b, 471–472.

[32] *celle qui n'est ni plus grande que zero, ni plus petite que zero, ni égale à zero*. Euler 1749, §3.

operations (taking logarithms, sines or cosines, for example) of numbers $M + N\sqrt{-1}$ also give rise to further numbers of the same form.[34] According to Euler, d'Alembert had recently proved the same thing but using arguments involving infinitely small quantities; this did not invalidate the proof for Euler but his own proof deliberately avoided such techniques.[35] The practical business of solving equations had led only to solutions that were real numbers or else numbers of the form $M \pm N\sqrt{-1}$; nevertheless for Euler and his contemporaries, the possible existence of other kinds of non-real numbers still needed to be carefully considered.

In his *Institutione calculi differentialis* of 1755, Euler included a chapter entitled 'De usu differentialium in investigandis radicibus realibus aequationum' ('The use of differentials in the investigation of real roots of equations').[36] Euler's arguments were essentially similar to those of Stirling and de Gua, identifying the existence of real or imaginary roots from the successive values of maxima and minima, but his exposition was more general and more lucid than some others that had preceded it. By probing the matter in greater detail Euler was able to refine the rule given by Stirling for cubic equations.[37] For the equation $x^3 - Ax^2 + Bx - C = 0$ to have three real roots, for instance, he claimed that we need not only $A^2 > 3B$, as proposed by Stirling, but also $\frac{1}{27}(A+f)^2(A-2f) < C < \frac{1}{27}(A-f)^2(A+2f)$ where f is the (positive) square root of $A^2 - 3B$. Euler derived similar rules for quartics also, but here the number of cases and sub-cases proliferated rapidly.[38]

Euler saw quite clearly that his method could not be applied to higher degree equations in general because of the difficulty of solving the derived equation. There are however, two special classes of equation where some useful information can be obtained. The first is the class of three-term equations, of the form $x^{m+n} + Ax^n + B = 0$, which had figured so prominently in the work of Cardano, Viète, and Harriot (see pages 11–12, 24–25, 32–33, 41).[39] The second is the class of equations where the derived equation is essentially quadratic. As an example Euler gave the equation $x^7 - 2x^5 + x^3 - a = 0$. The first derived equation is $7x^6 - 10x^4 + 3x^2 = 0$, which is essentially a quadratic in x^2 and can therefore be solved. Thus Euler in the mid-eighteenth century, with all the power of the calculus at his disposal, was forced to resort to the same special cases as his sixteenth-century predecessors, a sign of just how intractable the problem of detecting the existence of real roots was turning out to be.

---

33 *il paroit très vraisembable que toute racine imaginaire, quelque compliquée quelle soit, est toujours réductible à la forme* $M + N\sqrt{-1}$. [it seems very likely that every imaginary root, however complicated, is always reducible to the form $M + N\sqrt{-1}$.] Euler 1749, §64.

nous verrrons, qu'aucune opération ne nous sauroit écarter de cette forme [we see that no operation can take us away from this form] Euler 1749, §76.

34 Euler 1749, §78–§124.

35 See d'Alembert (1746) [1748], §II, Prop I.

36 Euler 1755, 523–547.

37 Euler 1755, 531–534.

38 Euler 1755, 534–540.

39 Euler 1755, 540–544.

Newton's rule for imaginary roots, 1760 onwards

By 1760, thanks to the work of de Gua and von Segner, Descartes' rule could be considered proved. A proof of Newton's rule, however, remained a *desideratum*, as Euler pointed out in the opening sentence of a paper that he wrote in the late 1760s, 'Nova criteria radices aequationum imaginarias dignoscendi' ('New criteria for discerning the imaginary roots of equations').[40] (By 1767 Euler was once again living in St Petersburg and publishing in the *Novi commentarii.*) Possibly he was inspired to take up the problem by the re-publication in Latin in 1761 of the papers by Maclaurin and Campbell in Johann Castillon's new and heavily footnoted edition of Newton's *Arithmetica universalis*.[41]

Euler began by observing what was by now well recognized: that all the criteria so far proposed for the existence of imaginary roots, including Newton's, were necessary but not sufficient. Where they indicated the existence of imaginary roots, such roots were sure to be found, but they might fail to indicate any at all even where *all* the roots are imaginary. Take for example, the equation

$$x^4 + 4x^3 - 8xx - 24x + 108 = 0.$$

All the known rules failed to identify the existence of any imaginary roots, yet all four roots of this equation are imaginary, as can be seen from the factorization

$$x^4 + 4x^3 - 8xx - 24x + 108 = (xx + 8x + 18)(xx - 4x + 6).$$

To improve upon this situation, Euler offered three principles (*principia*).

Principle 1. We can form an equation whose roots are the squares of the roots of the original equation. Euler did this by writing even powers of x on the left and odd powers on the right, then squaring each side and replacing xx by a new unknown, z. If all the roots (x) of the original equation are real then all the roots (z) of the new equation will be positive and its coefficients will have alternating signs. This criterion is not sufficient, however, to indicate the presence of imaginary roots. The equation $x^4 + 4x^3 - 8xx - 24x + 108 = 0$ given above, or indeed any equation of the form $x^4 + px^3 - qxx - rx + s = 0$ will give rise to an equation in z with alternating signs, but (as above) all its roots may be imaginary.

Principle 2. Suppose an equation

$$x^n + ax^{n-1} + bx^{n-2} + \cdots = 0$$

has roots $\alpha, \beta, \gamma, \ldots$. When all the roots are real we will have

$$(\alpha - \beta)^2 + (\alpha - \gamma)^2 + \cdots \geq 0,$$

[40] Euler 1768, E370.

[41] Castillon 1761, II, 61–109. The papers had already been published in Latin by Willem 'sGravesande in his earlier edition of the *Arithmetica universalis* in 1732, very soon after their original publication in English: 'sGravesande 1732, 298–344. There is indirect evidence (see pages 108–109) that Euler knew 'sGravesande's 1732 edition, but he may not then have thought Newton's rule worth taking up, or indeed may have assumed that Campbell and Maclaurin had already thoroughly dealt with it.

or

$$(\alpha^2 - 2\alpha\beta + \beta^2) + (\alpha^2 - 2\alpha\gamma + \gamma^2) + \cdots \geq 0.$$

Using the fact that the sum of the roots $(\alpha + \beta + \dots)$ is $-a$ and the sum of their products in pairs $(\alpha\beta + \alpha\gamma + \dots)$ is b, it is easy to obtain

$$(n-1)aa - 2nb \geq 0.$$

(which had been proved by Maclaurin and is one part of Newton's condition, though Euler did not comment on that at this point). This fails to guarantee, however, that every individual term of the form $(\alpha - \beta)^2$ is positive: once again we have a necessary but not sufficient condition for the existence of imaginary roots.

Principle 3. If an equation

$$x^n + ax^{n-1} + bx^{n-2} + \cdots = 0 \tag{10}$$

has all its roots real, we can form new equations of degree $n-1$ which will also have only real roots. Thus, for example,

$$nx^{n-1} + (n-1)ax^{n-2} + (n-3)bxn - 3 + \cdots = 0, \tag{11}$$

which Euler described as being formed from (10) by multiplying term-by-term by the arithmetic progression $n, n-1, n-2, \ldots$, and dividing by x. Or

$$ax^{n-1} + 2bx^{n-2} + 3cx^{n-3} + \cdots = 0, \tag{12}$$

formed by term-by-term multiplication of (10) by 0, 1, 2, 3, ... So far Euler appeared to be relying on a version of the theorem proved by Rolle and Maclaurin (see Theorem III above, page 91, and equations (8) and (9)). Unlike either of them, however, he also invoked calculus, pointing out that if $y = x^n + ax^{n-1} + bx^{n-2} + \dots$ then $\frac{dy}{dx} = nx^{n-1} + (n-1)ax^{n-2} + \dots$, so that (11) has $n-1$ real roots corresponding to the maxima or minima between the n real roots of (10). This argument does not, of course, explain why (12) also has $n-1$ real roots.

Another new equation with all its roots real can be obtained from (10) by putting $y = \dfrac{1}{x}$ to give

$$1 + ay + byy + 3cy^3 + \cdots = 0.$$

This too can be differentiated to form further equations with only real roots, for example,

$$a + 2by + 3cy^2 + \cdots = 0. \tag{13}$$

Indeed, we may continue in this way to find equations of any lower degree whose roots are all real. This was very similar to the approach Maclaurin had taken in 1730, except that Euler derived equations like (12) and (13) by differentiation and explained their properties by reference to curves, whereas Maclaurin had treated them from purely algebraic considerations.

Euler saw that the application of principle 2 to new equations like (12) and (13) produces several conditions for real roots. For a cubic equation $x^3+axx+bx+c=0$ to have all its roots real, for example, he found two necessary conditions

$$aa > 3b, \quad bb > 3ac,$$

while for the equation $x^6+ax^5+bx^4+cx^3+dx^2+ex+f=0$ he obtained

$$aa > \frac{12}{5}b, \quad bb > \frac{15}{8}ac, \quad cc > \frac{16}{9}bd, \quad dd > \frac{15}{8}ce, \quad ee > \frac{12}{5}df.$$

Only now did he acknowledge that these were the rules Newton had set out in his *Arithmetica universalis*.

The above work takes up the first half of Euler's paper. The remaining half consists of his attempts to refine these rules. To some extent he succeeded, finding, for example, that the exact criterion for a cubic equation $x^3+axx+bx+c=0$ to have all its roots real is

$$\left(\frac{a-\sqrt{aa-3b}}{3}\right)^3 < \frac{bb-3ac}{aa-3b} < \left(\frac{a+\sqrt{aa-3b}}{3}\right)^3,$$

a rule that includes both of those given earlier. For equations of higher degree, however, the calculations become laborious and after a while Euler could pursue them no further.

Thus, using a method very similar to Maclaurin's, Euler was able to demonstrate that Newton's rule was correct. Further, just as Maclaurin had done, he was able to offer more precise criteria, but he was still not able to solve the problem completely. The work of both Euler and Maclaurin suffered from the logical flaw of starting from equations whose roots were presumed real, and deriving criteria that must then hold. Where those criteria failed, one could be sure to find at least one pair of imaginary roots, but where they were satisfied, one could not be sure of anything at all.

Additional thoughts from Lagrange, 1769 and 1777

Some further thoughts on detecting imaginary roots were offered by Lagrange in 1769 in connection with his research on solving numerical equations. He wrote three papers on this subject: 'Sur la résolution des équations numériques' ('On the solution of equations with numerical coefficients'), and two later 'Additions'. These papers will be discussed in greater detail in Chapter 9 and are mentioned here only with respect to finding the number of imaginary roots.

In the first of the three papers, Lagrange suggested forming an equation whose roots are the squares of the differences of the roots of the proposed equation (we will return to his method of doing this later). He then argued that this new equation will have as many negative roots as there are pairs of imaginary roots in the proposed equation, since each imaginary pair $\alpha+\beta\sqrt{-1}$, $\alpha-\beta\sqrt{-1}$ gives rise to a difference $2\beta\sqrt{-1}$

whose square is $-4\beta^2$. In the second paper, Lagrange explored this idea further, and extracted useful information from the pattern of signs and sign changes in the equation for squares of differences. By looking at the sign of the final term, for example, he deduced that the number of real roots must belong to the sequence 1, 4, 5, 8, 9, 12, 13, ... if the sign was negative, or 2, 3, 6, 7, 10, 11, ... if it was positive. He hoped that by pushing the theory further it might be possible to determine the number of real roots exactly for any equation of any degree, but admitted that all methods devised so far fell short of that aim. Those of Newton and Maclaurin, he said, were insufficient, while those of Stirling and de Gua were impracticable.

Lagrange returned to the subject in the 1770s. We know from the records of the Berlin Academy that on 18 June 1772 he presented a paper entitled 'Recherches sur la maniere de déterminer le nombre des racines imaginaires qui peuvent se trouver dans les équations de tous les degrés' ('Researches on a method of determining the number of imaginary roots to be found in equations of any degree').[42] This paper was never published and its contents are unknown. Five years later, however, on 2 January 1777, he presented another paper with a very similar title, 'Recherches sur la détermination du nombre des racines imaginaires dans les équations litérales' ('Researches on determining the number of imaginary roots of literal equations'), and this later paper was published in the *Mémoires* for 1777 (printed 1779).

Lagrange began, as so often, by describing the historical background to the problem. He particularly commended Harriot, 'the learned English analyst' (*le savant Analyste Anglois*), as the first to have offered an algebraic proof of Cardano's condition for discerning whether a cubic equation has imaginary roots. Indeed, he repeated Harriot's entire proof from the *Praxis* (1631) together with some refinements of his own.[43] Lagrange noted that Harriot had not pushed such researches beyond cubic equations, and nor had anyone else until Newton offered his rule in the *Arithmetica universalis*. But Newton's rule, he complained, was clearly imperfect even with the additions of Maclaurin and Campbell. Lagrange, as usual, saw the problem clearly: all of these writers began by assuming that all the roots were real, and therefore arrived at conditions that were necessary for all the roots to be real, but not sufficient. Lagrange therefore proposed a different approach: for any proposed equation to produce a related equation in which the number of negative roots would correspond exactly to the number of imaginary roots of the proposed equation. An equation whose roots are the squares of the differences of the roots of the original equation would be just such an equation, as he had already suggested in 'Sur la résolution des équations numériques' and its 'Additions'. The problem with such an approach, as Lagrange recognised, is that the rule for determining the number of negative roots of an equation works only if one knows *a priori* that all the roots are real, precisely the problem one is concerned with. (Another problem is that the new equation will be of higher degree than the original, but this matters less because one is not concerned with solving it, only with reading off variations in sign.) Lagrange was able to work out his method for cubics and quartics,

[42]The presentation is recorded in the Academy *Registre*, BBAW MS I–IV–32, f. 113v.

[43]Harriot 1631, 80–83; Lagrange 1779, 112–114.

and to obtain some partial results for higher degree equations, but by the end of his paper his original problem still remained unresolved.

Thus, from as early as 1545, there were numerous attempts to discern the nature of the roots of equations, whether positive, negative, or imaginary, but by the late eighteenth century there was still no infallible rule for determining the number of real or imaginary roots for equations of degree higher than four. It was clear that the conditions given by Newton, Maclaurin, Campbell, and Euler were necessary for the existence of imaginary roots but not sufficient. It was also clear that the calculation of more precise conditions becomes seriously more difficult as the degree of the equation increases. There was not even any reason to hope that general conditions for sufficiency existed. With regard to this apparently simple problem, Lagrange was correct: there had been little practical advance since the time of Cardano.

Chapter 5
Roots as sums of radicals

In 1545 Cardano had written at some length about the number of positive or negative roots one could expect to find for a given cubic or quartic equation (see pages 14–16). He wrote very much more briefly about the *form* those roots could take (page 16). In his view a solution to a quadratic equation was the sum of a rational and a square root while a solution to a cubic equation was the sum of a rational and two cube roots. He did not explicitly discuss the structure of the roots of quartic equations, which from experience he knew to be rather more complicated (for an example see page 14). His only comment on equations of higher degree was that a fifth root, for example, could satisfy only an equation of the simplest kind, a fifth power equal to a number; conversely, such equations could not be satisfied by a sum of two or more such roots.

From now on, to avoid confusion between roots of numbers and roots of equations, we will use the term 'radicals' to describe square, cube, and all higher roots of integers or rational numbers. These are central to the content of this chapter.

Until the early years of the eighteenth century, no other author explicitly considered the form that roots of equations might take. When Dulaurens in 1667 solved some special equations of degrees 5, 7, and 11 they turned out to have roots composed of sums of pairs of radicals of degrees 5, 7, and 11, respectively, but Dulaurens did not comment on it. In Paris in 1675 Leibniz and Tschirnhaus briefly pursued the idea of roots as sums (or other expressions) composed of radicals, but Leibniz complained of the labour involved in trying to eliminate the radical signs, so the idea came to nothing and was never published (see pages 64–65). In the spring of 1707, however, two papers on equations were published in the *Philosophical Transactions of the Royal Society*, the first by John Colson, the second by Abraham de Moivre. Both introduced new conjectures about the structure of roots of equations. De Moivre's paper in particular was the mathematical starting point for the developments outlined in this chapter, and was quoted frequently by later writers.

In this chapter we will first discuss the papers of Colson and de Moivre. We will then examine the way the ideas presented in them were taken up first by Euler, who was quick to spot their potential, and later also by Étienne Bezout. The consequence was that Euler and Bezout were independently but almost simultaneously able to develop an important new technique of equation-solving, which will be described in the final part of this chapter.

The papers of Colson and de Moivre, 1707

John Colson, born in 1680, entered Christ Church, Oxford, in 1699 but never took his degree. Ten years later he took up a teaching post at the new mathematical school at Rochester in Kent. In 1739 he became a lecturer at Cambridge, and later that year became the fifth Lucasian Professor. Despite ending up in such a prestigious post, Colson's mathematical output over his lifetime was of little significance. He was better known as a publisher and translator of mathematical texts than as an innovator, and his paper on equations of 1707 was one of only three original pieces of work that he published. Nevertheless, it contained one important new idea.

The first part of the paper is devoted to the rules for solving cubic and quartic equations. For cubic equations Colson first stated the solution formula, then gave several worked examples to show its use. Only after that did he offer a derivation of it. His method, for solving the general equation $z^3 = 3qz + 2r$, was to suppose that $z = a + b$, so $z^3 = 3abz + a^3 + b^3$. Comparing this identity with the proposed equation we have

$$q = ab \quad (\text{or } q^3 = a^3b^3)$$

and

$$2r = a^3 + b^3.$$

These equations are easily combined to give

$$2ra^3 = a^6 + q^3,$$

which is a quadratic equation in a^3 with solutions

$$a^3 = r + \sqrt{r^2 - q^3}, \tag{1}$$

$$b^3 = r - \sqrt{r^2 - q^3}. \tag{2}$$

There was nothing new or remarkable in this (see, for example, similar derivations by Hudde and Dulaurens, pages 54–55 and 57–58). At this point, however, Colson observed that any quantity has three cube roots,[1] and that the cube roots of unity are 1, $-\frac{1}{2} + \frac{1}{2}\sqrt{-3}$, $-\frac{1}{2} - \frac{1}{2}\sqrt{-3}$. Equations (1) and (2) therefore yield three possible values for a and three for b. Thus there are potentially nine possible values of $z = a + b$. Colson tested out the various combinations, and found that the juxtapositions that satisfy the original equation are

$$\sqrt[3]{r + \sqrt{r^2 - q^3}} + \sqrt[3]{r - \sqrt{r^2 - q^3}},$$

$$\frac{-1 + \sqrt{-3}}{2} \times \sqrt[3]{r + \sqrt{r^2 - q^3}} + \frac{-1 - \sqrt{-3}}{2} \times \sqrt[3]{r - \sqrt{r^2 - q^3}},$$

$$\frac{-1 - \sqrt{-3}}{2} \times \sqrt[3]{r + \sqrt{r^2 - q^3}} + \frac{-1 + \sqrt{-3}}{2} \times \sqrt[3]{r - \sqrt{r^2 - q^3}}.$$

Thus he had found not just one root, as most of his predecessors had been satisfied to do, but all three roots of the original cubic.[2]

[1] *Cujusvis enim quantitatis Radix Cubica triplex erit* [The cube root of any quantity is threefold]. Colson 1707, 2356.

[2] Leibniz had asserted privately to Huygens that Cardano's rule could produce all the roots of a cubic, but had not explained how. See Chapter 1, note 19.

Colson also derived formulae for the roots of a quartic equation. Thus, according to Colson, the four solutions of $x^4 = 4px^3 + (2q - 4p^2)x^2 + (8r - 4pq)x + (4s - q^2)$ are

$$x = p - a \pm \sqrt{p^2 + q - a^2 - \frac{2r}{a}},$$

$$x = p + a \pm \sqrt{p^2 + q - a^2 + \frac{2r}{a}},$$

where a^2 is a root of the equation

$$a^6 = (p^2 + q)a^4 - (2pr + s)a^2 + r^2.$$

The fact that a cubic equation has three roots and a quartic equation has four had been recognized for at least a century but Colson was the first to give explicit formulae for each root. His paper ends with geometric constructions which need not concern us here.

Abraham de Moivre, who was just three years older than Colson, had arrived in England from France as a Protestant refugee shortly after the revocation of the Edict of Nantes in 1685. In mathematical and scientific circles he became highly respected, not least by Newton. He would undoubtedly have made a better candidate than Colson for the Lucasian chair in 1739, but because of his nationality was never able to obtain an academic position in England and instead eked out a living by private tutoring.

De Moivre's treatment of some special equations of third, fifth, seventh, ninth, or higher odd degree was published in the *Philosophical Transactions* immediately after Colson's paper. His exposition took the form of a claim followed by several worked examples. His paper opens with the following equation:

$$ny + \frac{nn-1}{2 \times 3} ny^3 + \frac{nn-1}{2 \times 3} \times \frac{nn-9}{4 \times 5} ny^5 + \frac{nn-1}{2 \times 3} \times \frac{nn-9}{4 \times 5} \times \frac{nn-25}{6 \times 7} ny^7 + \text{\&c.} = a \tag{3}$$

If n is an odd number, the series will terminate to give a finite equation. De Moivre claimed that a root of such an equation is

$$y = \tfrac{1}{2} \sqrt[n]{\sqrt{1 + aa} + a} - \frac{\frac{1}{2}}{\sqrt[n]{\sqrt{1 + aa} + a}}$$

or, equivalently,

$$y = \tfrac{1}{2} \sqrt[n]{\sqrt{1 + aa} + a} - \tfrac{1}{2} \sqrt[n]{\sqrt{1 + aa} - a}.$$

Thus, for instance, a solution to the equation

$$5y + 20y^3 + 16y^5 = 4$$

is

$$y = \tfrac{1}{2} \sqrt[5]{\sqrt{17} + 4} - \frac{\frac{1}{2}}{\sqrt[5]{\sqrt{17} + 4}}.$$

(2370)

Hic venit Observandum quod loco Radicis generalis, non incommode sumeretur $y = \frac{1}{2}\sqrt[n]{2a} - \frac{\frac{1}{2}}{\sqrt[n]{2a}}$, si quando numerus a respectu unitatis, sit satis magnus, ut si Æquatio fuerit $5y + 20y^3 + 16y^5 = 682$, erit Log. $2a = 3.1348143$, cujus pars quinta 0.6269628, & huic respondens numerus 4.236. Complementi autem Arithmetici 9.3730372 numerus est 0.236 & horum numerorum semidifferentia $2 = y$.

Atqui præterea, si in Æquatione præcedenti signa alternatim sint affirmantia & negantia, vel quod eodem redit, si series obvenerit hujus modi

$$ny + \frac{1-nn}{2\times3}ny^3 + \frac{1-nn}{2\times3}\times\frac{9-nn}{4\times5}ny^5 + \frac{1-nn}{2\times3}\times\frac{9-nn}{4\times5}\times\frac{25-nn}{6\times7}ny^7, \text{ \&c.} = a$$

erit hujus Radix

$$(1)\quad y = \tfrac{1}{2}\sqrt[n]{a+\sqrt{aa-1}} + \frac{\frac{1}{2}}{\sqrt[n]{a+\sqrt{aa-1}}}$$

$$\text{vel }(2)\quad y = \tfrac{1}{2}\sqrt[n]{a+\sqrt{aa-1}} + \tfrac{1}{2}\sqrt[n]{a-\sqrt{aa-1}}$$

$$\text{vel }(3)\quad y = \frac{\frac{1}{2}}{\sqrt[n]{a-\sqrt{aa-1}}} + \tfrac{1}{2}\sqrt[n]{a-\sqrt{aa-1}}$$

$$\text{vel }(4)\quad y = \frac{\frac{1}{2}}{\sqrt[n]{a-\sqrt{aa-1}}} + \frac{\frac{1}{2}}{\sqrt[n]{a+\sqrt{aa-1}}}$$

Hic autem Notandum, quod si $\frac{n-1}{2}$ numerus exciterit impar, Radicis inventæ signum in ei contrarium permutandum est.

Pro-

Some equations that can be solved by radicals, from de Moivre (1707).

De Moivre evaluated this using logarithms to obtain $y = 0.4313$. Next he gave similar rules for equation (3) when the signs alternate. Thus, he claimed that a root of

$$5y - 20y^3 + 16y^5 = \frac{61}{64} \tag{4}$$

is

$$y = \tfrac{1}{2}\sqrt[5]{\tfrac{61}{64} + \sqrt{\tfrac{-375}{4096}}} - \tfrac{1}{2}\sqrt[5]{\tfrac{61}{64} - \sqrt{\tfrac{-375}{4096}}}.$$

Note that equation (4) is of the special kind that Dulaurens had also been able to solve in 1667 (see pages 57–58).

At the beginning of the eighteenth century the astute reader, familiar with Newton's multiple angle equations published several years earlier by John Wallis,[3] would have recognized that equation (3) with alternating signs relates $\sin n\theta$ (represented by $\pm a$) to $\sin\theta$ (represented by y). De Moivre knew this. The link to angle division came at the end of his paper, where he remarked that using tables of sines one can extract a positive real root (*proba et possibilis*) of the seemingly 'impossible' binomial $\frac{61}{64} + \sqrt{\frac{-375}{4096}}$ that arises in the solution of (4). He explained how to do this by first calculating $\frac{61}{64} = 0.95312$ (though in the published version this is misprinted as 0.95112). He then stated that 0.95312 is $\sin 72^\circ\, 23'$; and that one fifth of this angle is $14^\circ\, 28'$, whose sine is 0.24981, which is very close to $\frac{1}{4}$. He did not say how to calculate the imaginary part of the fifth root, but one can repeat a similar process or, more simply, use the fact that $1 - (\frac{1}{4})^2 = (\frac{\sqrt{15}}{4})^2$. Thus, de Moivre's estimate of the fifth root was

$$\tfrac{1}{4} + \tfrac{1}{4}\sqrt{-15}.$$

This was the first hint of what later came to be called de Moivre's theorem, for de Moivre was clearly using the relationship

$$\sqrt[n]{(\sin n\theta + \mathrm{i}\cos n\theta)} = \sin\theta + \mathrm{i}\cos\theta.$$

The background and full development of the ideas in this paper of 1707 were not published until 1730 in de Moivre's *Miscellaneae*.[4] For our purposes, however, what matters is that by 1707 de Moivre had put forward a general class of equations of odd degree for which there are known solutions. We can recognize these as angle division equations, the lower cases of which were also known to Viète, Briggs, and Dulaurens. De Moivre's achievement was to use Newton's multiple angle formulae to describe this class for any odd degree.

Euler's conjecture, 1733

In 1732, Willem 'sGravesande republished the papers by Colson and de Moivre as appendices to his edition of Newton's *Arithmetica universalis*.[5] It was probably there

[3] Wallis 1685, 341–342.
[4] De Moivre 1730, 13–26.
[5] 'sGravesande 1732, 258–273.

rather than in the *Philosophical Transactions* of 1707 (the year Euler was born) that Euler first read de Moivre's paper, and probably Colson's too. The following year, 1733, Euler presented to the St Petersburg Academy a paper entitled 'De formis radicum aequationum cuiusque ordinis coniectatio' ('A conjecture on the form of roots of equations of any degree') in which he specifically referred to de Moivre's findings. As was common throughout the eighteenth century, there was significant delay between the original presentation of the paper and its publication in the proceedings of the Academy, the *Commentarii Academiae Scientiarum Petropolitanae*.[6] The volume for 1732–33, including Euler's paper, was eventually printed in 1738.

Euler began with the observation that Lagrange was to echo later: that rules for cubic and quartic equations had been found at the beginning of investigations into such matters but despite many advances in analysis since that time there had been no progress with equations of higher degree. He went on to comment that it was 'easily seen' (*facile perspicitur*) that solving an equation of any degree will depend on the ability to solve all equations of lower degree, just as solving a cubic requires the solution of a quadratic, and solving a quartic requires the solution of a cubic. The investigation that followed was typical of Euler both in the structure of the argument and in the clarity of his writing. He began with cubics, which were relatively easy to examine; then moved on to quartics; then returned to the simple case of quadratics to check that his findings held there too; then finally extended his investigations to equations of degree five or higher.

As an example of a general cubic equation lacking its square term Euler took $x^3 = ax + b$. He noted that a root of this equation takes the form $x = \sqrt[3]{A} + \sqrt[3]{B}$, where A and B are in turn roots of a quadratic equation $z^2 = \alpha z - \beta$. We can write down this quadratic equation immediately if we can determine $\alpha = A + B$ and $\beta = AB$ in terms of a and b, the coefficients of the proposed equation. Substituting $x = \sqrt[3]{A} + \sqrt[3]{B}$ back into the original equation Euler found that $A + B = b$ and $AB = a^3/27$, so that the required quadratic equation is

$$z^2 = bz - \frac{a^3}{27}.$$

As had been pointed out by Colson, however, and as was by now well known, there are three possible values for the cube root of A, namely, one that we may write as $\sqrt[3]{A}$ but also $\mu\sqrt[3]{A}$ and $\nu\sqrt[3]{A}$ where μ and ν are the two cube roots of unity not equal to 1; and similarly for B. Constructing sums of pairs therefore leads to nine possibilities for x. The additional requirement that the product of the pairs must equal β, however, reduces the nine possibilities to three, namely, $x = \sqrt[3]{A} + \sqrt[3]{B}$, $x = \mu\sqrt[3]{A} + \nu\sqrt[3]{B}$, and $x = \nu\sqrt[3]{A} + \mu\sqrt[3]{B}$. These, of course, were precisely the possibilities given by Colson in 1707. Euler did not mention Colson but it is very likely that he had seen his paper since he had certainly read de Moivre's, which followed immediately after it.

[6]By 1750 the publication delay at St Petersburg was up to ten years. The backlog was published in two final volumes, 13 and 14, of the *Commentarii*, which was then replaced by the *Novi commentarii* in 1751.

Moving on to quartic equations, Euler commented that there are several ways to solve such equations by finding and solving an intermediate cubic, but that his purpose here was to show a method that might be capable of generalization to equations of higher degree. Suppose, therefore, that we have a quartic without a cube term, of the form $x^4 = ax^2 + bx + c$. Suppose too that it has a root expressible as $x = \sqrt{A} + \sqrt{B} + \sqrt{C}$ where A, B, and C are roots of a cubic equation $z^3 = \alpha z^2 - \beta z + \gamma$. Proceeding as for the cubic case, Euler discovered that $A + B + C = \frac{a}{2}$, $AB + AC + BC = \frac{c}{4} - \frac{a^2}{16}$, and $ABC = \frac{b^2}{64}$ and so the required cubic equation is

$$z^3 = \frac{a}{2}z^2 - \frac{4c - a^2}{16}z + \frac{b^2}{64}.$$

The three roots of this equation are thus A, B, and C, and Euler took one root of the original equation to be $x = \sqrt{A} + \sqrt{B} + \sqrt{C}$, without commenting on the ambiguity of the radical sign. He added that the other three roots will be $x = \sqrt{A} - \sqrt{B} - \sqrt{C}$, $x = \sqrt{B} - \sqrt{A} - \sqrt{C}$, and $x = \sqrt{C} - \sqrt{A} - \sqrt{B}$; these are correct but Euler did not show how he had arrived at these particular combinations of sign. Next he put $A = \sqrt{E}$, $B = \sqrt{F}$, and $C = \sqrt{G}$, again without commenting on the ambiguity of sign. Using this trick, however, he now had his roots in the form $\pm\sqrt[4]{E} \pm \sqrt[4]{F} \pm \sqrt[4]{G}$ taking appropriate combinations of sign; that is, he was able to claim that the roots of a quartic equation can be expressed by a formula analogous to that for the roots of a cubic.

Before going on to higher degree equations Euler checked his results on the quadratic case. A quadratic equation without a linear term has the simple form $x^2 = a$. It is solved, Euler argued, by means of an equation one degree lower, that is $z = a$ whose only root is a. The solutions of the original equation are then $x = \sqrt{a}$ or $x = -\sqrt{a}$. Euler called an equation of the form $z = a$ (for quadratics) or $z^2 = \alpha z - \beta$ (for cubics) a 'resolvent equation' (*aequatio resoluentis*), the first use of this term for an equation of lower degree than the original, by means of which the original can be solved.

Thus Euler could claim that, taking appropriate values of the radical in each case, we have the following results. For quadratics, if the root of the resolvent is A, we have $x = \sqrt{A}$; for cubics, if the roots of the resolvent are A and B then $x = \sqrt[3]{A} + \sqrt[3]{B}$; and for quartics, if the roots of the resolvent are A, B, and C then $x = \sqrt[4]{A} + \sqrt[4]{B} + \sqrt[4]{C}$.

On the basis of this evidence, Euler was led to the conjecture suggested in the title of his paper: that similar results must also hold for equations of higher degree. Thus the roots of the general quintic $x^5 = ax^3 + bx^2 + cx + d$ will be of the form $x = \sqrt[5]{A} + \sqrt[5]{B} + \sqrt[5]{C} + \sqrt[5]{D}$ where A, B, C, D, are the roots of a quartic equation $z^4 = \alpha z^3 - \beta z^2 + \gamma z - \delta = 0$, and so on. There is an obvious generalization to any equation of degree n lacking a term in x^{n-1}. Unfortunately, Euler was forced to admit that for equations of degree higher than four he had so far been unable to construct the coefficients of the resolvent. Nevertheless, he claimed that some partial results confirmed his conjecture.

The partial results of which Euler spoke were for equations whose resolvent was of the special form $z^{n-1} = \alpha z^{n-2} - \beta z^{n-3}$, or $z^2 = \alpha z - \beta$. Such equations, said Euler, were precisely those identified and solved by de Moivre in his paper of 1707. We can see this, he argued, because the resolvent $z^2 = \alpha z - \beta$ has only two possible roots A and B, and the roots of the proposed equation are then simply $\sqrt[n]{A} + \sqrt[n]{B}$, as claimed by de Moivre. In fact, as in the cubic case already examined, the roots may be actually $\mu\sqrt[n]{A} + \nu\sqrt[n]{B}$ where μ and ν are n^{th} roots of unity and $\mu\nu = 1$.

Euler had run into difficulty trying to construct the resolvent of a fifth-degree equation. Towards the end of his paper he tackled the converse problem of constructing an equation from its resolvent but was forced to abandon that too. In short, his general method for solving higher degree equations without a second term did not seem to be working out very well. What mattered most to those who followed him however, was Euler's suggestion that the roots of an equation of degree n could be written as a sum of up to $n - 1$ radicals of degree n.

The papers of Euler and Bezout, 1764

Euler does not seem to have followed up his idea about roots as sums of radicals until about twenty years later, by which time he had left the Academy of St Petersburg for that of Berlin. On 3 May 1753 he presented to the Berlin Academy a paper entitled 'De resolutione aequationum cuisvis gradus' ('On the solution of equations of any degree').[7] For some reason the paper was not published in the Berlin *Mémoires* but was communicated to the St Petersburg Academy in October 1759. It was subsequently published in the *Novi commentarii* for 1762–63, which was printed in 1764. In that same year, 1764, the Paris Academy published its *Mémoires* for 1762, which included a paper by Étienne Bezout on a precisely similar subject. Bezout had read Euler's conjecture of 1733 but was unaware of his subsequent work until their two papers were published almost simultaneously in 1764. Since Euler's findings had been developed about ten years earlier than Bezout's, we will begin with those.

Euler began by remarking that there was still no general rule for solving equations of degree higher than four, but that de Moivre had identified certain special equations, which could not be factorized but which could nevertheless be solved. Further, he noted once again that equations of degree one (linear equations) can be solved without any extraction of roots, that quadratic equations can be solved using square roots, cubic equations using square and cube roots, and quartic equations using at most fourth roots. It was therefore reasonable to suppose that an equation of any degree could be solved by means of radicals of that degree and lower.

More precisely he referred back to his conjecture of 1733 (though now using slightly different notation). Thus he supposed that an equation

$$x^n + Ax^{n-2} + Bx^{n-3} + \cdots = 0$$

[7]The presentation was recorded in the Academy's *Registre* for 1746–66, BBAW MS I–IV–31, f. 108. The secretary, however, mistakenly wrote 'generis' ('kind') instead of 'gradus' ('degree'), leading to some confusion later as to the correct title of the paper, see BBAW MS C.5, f. 3v. A manuscript copy of the paper, now known as E282, is held in the Academy archives as BBAW MS I–M 122, C.7, ff. 222–229; there, as in the published version, the final word is 'gradus'.

has roots of the form

$$x = \sqrt[n]{\alpha} + \sqrt[n]{\beta} + \sqrt[n]{\gamma} + \cdots \tag{5}$$

where $\alpha, \beta, \gamma, \ldots$ are the $n-1$ roots of a resolvent equation one degree lower:

$$y^{n-1} + \mathfrak{A}y^{n-2} + \mathfrak{B}y^{n-3} + \cdots = 0.$$

Now, however, Euler commented that there were certain troublesome aspects about the form suggested in (5). The main problem is that it does not define a root x unambiguously because each expression $\sqrt[n]{\alpha}$ has n different values. This is because $\sqrt[n]{1}$ itself has n values, which Euler denoted by $1, \mathfrak{a}, \mathfrak{b}, \mathfrak{c}, \ldots$, so that for $\sqrt[n]{\alpha}$ we may substitute any of $\mathfrak{a}\sqrt[n]{\alpha}, \mathfrak{b}\sqrt[n]{\alpha}, \mathfrak{c}\sqrt[n]{\alpha}, \ldots$. If we allow all combinations of $1, \mathfrak{a}, \mathfrak{b}, \mathfrak{c}, \ldots$ with $\sqrt[n]{\alpha}, \sqrt[n]{\beta}, \sqrt[n]{\gamma}, \ldots$ we end up with far too many possibilities for x, which should have no more than n distinct values. It is clear, therefore, that the combinations must be restricted in some way so as to give only the true roots of the proposed equation. For equations of degree 3 or 4 Euler knew the rules for doing this. For cubics, for example, he used only those pairs of $1, \mu, \nu$ (cube roots of unity) whose product was 1. For quartics, he had also arrived, presumably by trial and error, at the correct combinations of $+$ and $-$ in his sums of $\sqrt[4]{E}$, $\sqrt[4]{F}$, and $\sqrt[4]{G}$. The equivalent rules for equations of higher degree, however, were not known. Euler now hoped to remove this inconvenience altogether by proposing a different form for the roots.

First he observed that if $\mathfrak{a}$ is any n^{th} root of unity then so are $\mathfrak{a}^2, \mathfrak{a}^3, \ldots, \mathfrak{a}^{n-1}$. He seems to have assumed that these would be distinct, which is not the case unless $\mathfrak{a}$ is what is now called a primitive root.[8] Further, he assumed that any one of the roots would, by repeated multiplication, generate all the rest.[9] These assumptions hold only if n is prime, for only then is it true that any root (except 1) generates all the others, a caveat that should be borne in mind in following the remainder of Euler's argument.

These observations led Euler to suspect that a similar pattern might hold in an expression like (5) for the root of an equation: that is, given one of the radicals, all the others would be powers of it. But then in order to retain $n-1$ unknown quantities, as in (5), each of the radicals, said Euler, must be multiplied by an arbitrary coefficient. Thus it seemed to him highly probable that the root x must take the form[10]

$$x = \mathfrak{A}\sqrt[n]{v} + \mathfrak{B}\sqrt[n]{v^2} + \mathfrak{C}\sqrt[n]{v^3} + \mathfrak{D}\sqrt[n]{v^4} + \cdots$$

---

[8] A primitive root of unity is one whose powers generate all the other roots. Amongst the fourth roots of unity, for example, i and $-$i are primitive roots, but 1 and -1 are not.

[9] *Ita si post vnitatem, quae semper primum locum tenere censenda est, a littera* $\mathfrak{a}$ *incipiamus, valores formulae* $\sqrt[n]{1}$ *erunt* $1, \mathfrak{a}, \mathfrak{a}^2, \mathfrak{a}^3, \mathfrak{a}^4 \ldots \mathfrak{a}^{n-1}$ *quorum numerus est n; plures enim occurrere nequeunt, cum fit* $\mathfrak{a}^n = 1, \mathfrak{a}^{n+1} = \mathfrak{a}, \mathfrak{a}^{n+2} = \mathfrak{a}^2$ *etc. similique modo res se habebit, si post vnitatem a quauis alia littera* $\mathfrak{b}$, *vel* $\mathfrak{c}$, *vel* $\mathfrak{d}$ *etc. incipiamus.* [Thus if after unity, which must always be thought to take the first place, we begin from the letter $\mathfrak{a}$, the values of $\sqrt[n]{1}$ will be $1, \mathfrak{a}, \mathfrak{a}^2, \mathfrak{a}^3, \mathfrak{a}^4 \ldots \mathfrak{a}^{n-1}$ which are n in number; for there cannot be more, since $\mathfrak{a}^n = 1, \mathfrak{a}^{n+1} = \mathfrak{a}, \mathfrak{a}^{n+2} = \mathfrak{a}^2$ etc. This will come about in a similar way if after unity we begin from any other letter $\mathfrak{b}$, or $\mathfrak{c}$, or $\mathfrak{d}$ etc.] Euler (1762) [1764], §7.

[10] *maxime probabile videtur radicem quamlibet huius aequationis ita exprimi* [it seems highly probable that any root of this equation can be expressed thus] Euler (1762) [1764], §8.

where v is some as yet unspecified quantity and $\mathfrak{A}$, $\mathfrak{B}$, $\mathfrak{C}$, ... are rational quantities that do not involve n^{th} roots of v. Since $\sqrt[n]{v^{n+1}} = v\sqrt[n]{v}$, $\sqrt[n]{v^{n+2}} = v\sqrt[n]{v^2}$, and so on, the above expression, like (5), contains at most $n-1$ summands, which helped to confirm for Euler that this new approach was correct.

Until now, the original equation had been assumed to be devoid of a term in x^{n-1}. If it contains the term Δx^{n-1}, however, the solution is adapted simply by adding an appropriate constant $w = -\frac{1}{n}\Delta$, thus

$$x = w + \mathfrak{A}\sqrt[n]{v} + \mathfrak{B}\sqrt[n]{v^2} + \mathfrak{C}\sqrt[n]{v^3} + \mathfrak{D}\sqrt[n]{v^4} + \cdots. \tag{6}$$

Euler thought that the heart of the whole problem was contained in (6). To see that it displays the roots without ambiguity, he argued, recall the principle that any root of unity can be combined with $\sqrt[n]{v}$. Thus we can replace $\sqrt[n]{v}$ by any of $\mathfrak{a}\sqrt[n]{v}$, $\mathfrak{b}\sqrt[n]{v}$, $\mathfrak{c}\sqrt[n]{v}$, ... Further, if we combine $\sqrt[n]{v}$ with $\mathfrak{a}$, then instead of $\sqrt[n]{v^2}$, $\sqrt[n]{v^3}$, $\sqrt[n]{v^4}$, ... we must write $\mathfrak{a}^2\sqrt[n]{v^2}$, $\mathfrak{a}^3\sqrt[n]{v^3}$, $\mathfrak{a}^4\sqrt[n]{v^4}$, ... The constant w is consistent with this pattern since it can be written as $w\mathfrak{a}^0\sqrt[n]{v^0}$.

Thus, by incorporating each root of unity in turn, expression (6) delivers all n roots of the original equation

$$\begin{aligned}
x &= w + \mathfrak{A}\mathfrak{a}\sqrt[n]{v} + \mathfrak{B}\mathfrak{a}^2\sqrt[n]{v^2} + \mathfrak{C}\mathfrak{a}^3\sqrt[n]{v^3} + \cdots + \mathfrak{D}\mathfrak{a}^{n-1}\sqrt[n]{v^{n-1}},\\
x &= w + \mathfrak{A}\mathfrak{b}\sqrt[n]{v} + \mathfrak{B}\mathfrak{b}^2\sqrt[n]{v^2} + \mathfrak{C}\mathfrak{b}^3\sqrt[n]{v^3} + \cdots + \mathfrak{D}\mathfrak{b}^{n-1}\sqrt[n]{v^{n-1}},\\
x &= w + \mathfrak{A}\mathfrak{c}\sqrt[n]{v} + \mathfrak{B}\mathfrak{c}^2\sqrt[n]{v^2} + \mathfrak{C}\mathfrak{c}^3\sqrt[n]{v^3} + \cdots + \mathfrak{D}\mathfrak{c}^{n-1}\sqrt[n]{v^{n-1}},\\
&\dots.
\end{aligned}$$

Euler ended this part of his exposition by saying once again that it seemed to him highly probable that he had discovered the correct form of the roots. To be certain of it, nothing more was required than to show how to find $\mathfrak{A}$, $\mathfrak{B}$, $\mathfrak{C}$, ... and v for any given equation. He had not so far been able to ascertain the rules for doing so, but nevertheless thought that what he had given so far would shed considerable light on the matter of solving equations.

In the second part of his paper (§15–§46), Euler explored the converse question: given a root of the form indicated in (6) can we find a polynomial equation with rational coefficients that it must satisfy? If we can, he claimed, it will not only confirm the conjecture but will also give us a class of solvable equations. However, even the simplest root $x = w + \mathfrak{A}\sqrt[n]{v}$, he noted, gives rise to an equation of degree n, and adding in further summands can only increase the degree of the equation. In fact a root of the form (6) contains $n-1$ arbitrary quantities $\mathfrak{A}$, $\mathfrak{B}$, $\mathfrak{C}$, ... from which $n-1$ others, A, B, C, ... must be determined, a problem he knew to be in general of considerable difficulty.

Including v as well gave Euler not $n-1$ but n unknown quantities, but he argued that one of them could always be chosen at will. For $n = 2$ and $n = 3$ he put $\mathfrak{A} = 1$, that is, he assumed roots of the form $x = \sqrt{v}$ and $x = \sqrt[3]{v} + \mathfrak{B}\sqrt[3]{v^2}$, respectively; while for

$n = 4$ he put $\mathfrak{B} = 1$, that is, he assumed a root of the form $x = \mathfrak{A}\sqrt[4]{v} + \sqrt[4]{v^2} + \mathfrak{C}\sqrt[4]{v^3}$. In each case he was able to show that the calculations led to the expected quadratic, cubic, or quartic equation (§20–§29). For $n = 5$, however, he once again ran into seemingly insuperable difficulties, ending up with equations of the fifth degree in $\mathfrak{A}$, $\mathfrak{B}$, $\mathfrak{C}$, and $\mathfrak{D}$. Even assigning an arbitrary value to one of them, Euler could see no way of eliminating the rest to find an equation for v. Nevertheless, he still hoped that if matters were handled correctly it would be possible to arrive at a solvable equation for v of degree four.[11]

In the meantime, there were certain special cases that Euler could deal with. One was the trivial case where $\mathfrak{B} = \mathfrak{C} = \mathfrak{D} = 0$ and the required equation is simply $x^5 = D$. He could also handle cases where any two of $\mathfrak{A}$, $\mathfrak{B}$, $\mathfrak{C}$, $\mathfrak{D}$ are zero. If $\mathfrak{C} = \mathfrak{D} = 0$, for example, the equation that arises for x is of the form

$$x^5 - 5Pxx + 5Qx + \left(\frac{QQ}{P} + \frac{P^3}{Q}\right) = 0,$$

where $P = \mathfrak{A}\mathfrak{B}^2 v$ and $Q = \mathfrak{A}^3\mathfrak{B}v^2$. This has roots $\alpha\sqrt[5]{\frac{QQ}{P}} + \alpha^2\sqrt[5]{\frac{P^3}{Q}}$ where α is a fifth root of unity. Euler noted that such equations are similar to those discovered by de Moivre, and observed that since they are irreducible their solutions are worth knowing. The case $\mathfrak{B} = \mathfrak{C} = 0$, which Euler addressed a few paragraphs later does in fact give rise to de Moivre's equations, as Euler saw and noted immediately.

The final paragraph of his paper contains two particular examples of irreducible equations that Euler could now solve. The second, deceptively simple in appearance, was $x^5 = 2625x + 16600$, one of whose roots is

$$\sqrt[5]{75(5+4\sqrt{10})} + \sqrt[5]{225(35+11\sqrt{10})} + \sqrt[5]{225(35-11\sqrt{10})} + \sqrt[5]{75(5-4\sqrt{10})}.$$

Euler's paper of 1764 clearly represents a considerable advance on his work of 1733. Then, he had conjectured that the roots of a polynomial of degree n are always of the form $x = \sqrt[n]{A} + \sqrt[n]{B} + \cdots$ with up to $n - 1$ summands. Now, by considering n^{th} roots of unity, he had arrived at a different hypothesis, which enabled him, as he thought, to list not just one but all n roots of an equation of degree n. Further, his list showed how the roots would relate to one another in a regular way. All the evidence Euler could collect, from equations of degree 2, 3, or 4, and a few special cases of higher degree, suggested to him that his new conjecture was correct. Unfortunately the difficulties in all other cases of finding roots from equations or, conversely, equations from roots, seemed to be insuperable, leading only to equations as bad as, or worse than, the original. Nevertheless, Euler was able to list a number of special classes of equations, in addition to those already identified by de Moivre in 1707, which could be solved.

[11] *Satis tuto autem suspicari licet, si haec eliminatio rite administretur, tandem ad aequationem quarti gradus perueniri posse, qua valor ipsius v definiatur.* [All this, moreover, sufficiently allows one to expect that if this elimination is correctly handled one can arrive at length at an equation of fourth degree, by which the value of v will be defined.] Euler (1762–63) [1764] §37.

Meanwhile, unknown to Euler, Bezout, also inspired by de Moivre's findings of 1707 and by Euler's paper of 1738, was following other approaches and arriving at similar conclusions.

Étienne Bezout was born in 1730 in Nemours, near Fontainebleau, some 80 kilometres south of Paris. His early reading of Euler had led him to publish a memoir on dynamics in 1756 and others on calculus in 1758. In that year he was elected an *adjoint* in mechanics at the Paris Academy, a form of association that meant he had to spend at least half a year in Paris. In 1763 he became teacher and examiner in mathematical sciences for the Gardes de la Marine, an elite corps of young men aged from about fifteen upwards, from amongst whom all French naval officers were drawn. Its bases were at Brest, Toulon, and Rochefort. Despite the travel this position must have entailed, Bezout wrote a six-part *Cours de mathématiques*, which was published from 1764 onwards (see pages 199–201), so during these years he must have been particularly busy. Nevertheless, it was during this time that he also did some of his most important early work on equations.

Bezout's first paper on the subject, his 'Mémoire sur plusieurs classes d'équations de tous les degrés qui admettent une solution algébrique' ('Memoir on several classes of equations of all degrees which allow an algebraic solution'), was presented to the Paris Academy in 1762 and published in the Academy's *Mémoires* in 1764. Thus, though Euler's paper had been considerably longer in gestation, both appeared in print in the same year. Bezout began with the kind of comment that was by now becoming commonplace amongst writers on equations: that with regard to solving equations of general degree there had been hardly any advance since the time of Descartes. Bezout, a careful reader of Euler, first summarized Euler's findings of 1738, and in particular his conjecture that a root of an equation of degree n is a sum of nth roots. Bezout observed, however, that Euler's introduction of fourth roots in his treatment of quartics was somewhat artificial, and that as yet the only support for Euler's hypothesis came from the equations identified by de Moivre in 1707 and those added by Euler in 1738. Bezout said that he himself had made many attempts to pursue similar ideas but with little success, but that other methods had led him to some useful results. Here he would present the method that seemed to him clearest.

Bezout's 'Problem I' (§13) was to illustrate his method as applied to cubic equations. Suppose, he said, that we wish to solve the equation

$$x^3 + px^2 + qx + r = 0. \tag{7}$$

Let us look for a transformation of the form

$$y = \frac{x+a}{x+b} \tag{8}$$

with suitable values of a and b, so that when this value of y is substituted into

$$y^3 + h = 0 \tag{9}$$

the resulting equation will be (7). In other words, we need to discover appropriate values of a, b, and h in (8) and (9) that will give the correct values of p, q, and r in (7).

Substituting (8) into (9) and comparing coefficients with (7) (and silently assuming $1 + h \neq 0$) gave Bezout

$$h = \frac{3a - p}{p - 3b},$$

and

$$a + b = \frac{pq - 9r}{pp - 3q},$$

$$ab = \frac{qq - 3pr}{pp - 3q}.$$

From the last two equations it is easy to write down the quadratic equation whose roots are a and b, and from its solutions one can find h (though Bezout did not explain how to deal with the ambiguity of the solutions a and b).

Now solving first (9) for y and then (8) for x gives

$$x = -\tfrac{1}{3}p + \tfrac{1}{3}\sqrt[3]{(3a - p)^2(3b - p)} + \tfrac{1}{3}\sqrt[3]{(3a - p)(3b - p)^2}.$$

(Here the old language of proportionals briefly reasserted itself: Bezout noted that the cube roots in this expression are the two mean proportionals between $(3a - p)$ and $(3b - p)$.)

When the original equation has no square term we have $p = 0$ and the results become simpler and more familiar. The quadratic equation whose roots are a and b becomes (as Bezout wrote it)

$$a^2 - \frac{3r}{q}a - \frac{q}{3} = 0.$$

Bezout, following Descartes, called this a 'reduced equation' (*réduite*); it is what Euler had called a 'resolvent equation' (*aequatio resoluens*). It led Bezout to

$$x = \sqrt[3]{a^2 b} + \sqrt[3]{ab^2}$$

as the solution to the original equation.

Bezout's 'Problem II' (§16), demonstrated how his method could be applied to an equation of any degree. Suppose we wish to solve

$$x^n + mx^{n-1} + px^{n-2} + qx^{n-3} + rx^{n-4} + \cdots + M = 0. \quad (10)$$

Bezout once again proposed a transformation

$$y = \frac{x + a}{x + b} \quad (11)$$

with appropriate values of a and b, so that this value of y substituted into

$$y^n + h = 0 \quad (12)$$

would give rise to (10). For simplicity Bezout this time assumed $m = 0$. With this condition, substitution of (11) into (12) leads to $h = -\frac{a}{b}$, and

$$\begin{aligned} x^n - n\frac{n-1}{2}abx^{n-2} \\ -n\frac{n-1}{2}\frac{n-2}{3}ab(a+b)x^{n-3} \\ -n\frac{n-1}{2}\frac{n-2}{3}\frac{n-3}{4}ab(a^2+ab+b^2)x^{n-4} \\ -n\frac{n-1}{2}\frac{n-2}{3}\frac{n-3}{4}\frac{n-4}{5}ab(a^3+a^2b+ab^2+b^3)x^{n-5} \\ -\cdots \\ -ab(a^{n-2}+a^{n-3}b+a^{n-4}b^2+\cdots+b^{n-2}) = 0. \end{aligned} \tag{13}$$

Comparing (13) with (10) we obtain n equations for $p, q, r, \ldots, M$ in terms of a and b. But Bezout noticed a short cut: it is 'easy to see' (*aisé de voir*), he observed, that all the coefficients in (13) can be expressed in terms of ab and $(a + b)$. Most readers might have found this less easy to see than Bezout did, but he was correct.

The quantities $a + b$ and ab can therefore be found in terms of p and q using just the second and third terms of (13) which give the equations

$$-n\frac{n-1}{2}ab = p,$$
$$-n\frac{n-1}{2}\frac{n-2}{3}ab(a+b) = q.$$

That is, a and b are the two roots of the quadratic equation that Bezout wrote as

$$a^2 - \frac{q}{\frac{n-2}{3}p}a - \frac{p}{n\frac{n-1}{2}} = 0.$$

(Here, of course, we must assume that $p \neq 0$; Bezout discussed the special cases $p = 0$ and $p = q = 0$ separately later.) From (11) we have

$$x = \frac{a - by}{y - 1}, \tag{14}$$

and so, substituting $y = \sqrt[n]{\frac{a}{b}}$, we have

$$\begin{aligned} x &= \frac{a\sqrt[n]{b} - b\sqrt[n]{a}}{\sqrt[n]{a} - \sqrt[n]{b}} \\ &= \sqrt[n]{a^{n-1}b} + \sqrt[n]{a^{n-2}b^2} + \sqrt[n]{a^{n-3}b^3} + \cdots + \sqrt[n]{ab^{n-1}}. \end{aligned} \tag{15}$$

Once again, as for the solution of a cubic, Bezout noted that this is a sum of $n - 1$ mean proportionals between a and b. Further, a and b are found from an equation of degree 2, which is easily solved. Thus equation (10) is solvable, and its solution is a special case of the form conjectured by Euler in 1733, whenever the coefficients take the restricted form set out in (13).

A modern understanding of the situation is that transformation (11), now known as a Möbius transformation, has the property that it preserves the set of circles (including straight lines) in the complex plane. Since the roots of (12) lie on a circle, transformation (11) is possible only if the roots of (10) also lie on a circle (or straight line). Clearly this is the case only when the coefficients of (10) satisfy certain highly restrictive conditions, as Bezout had discovered.

Bezout observed that so far the method had used only one possible root of $y^n = \dfrac{a}{b}$, but this equation has n roots, and there is no reason to select one rather than another; thus equation (14) will deliver *all* the roots of the original equation when the n possible values of y are substituted in turn. Bezout knew (quoting work of Roger Cotes)[12] that the equation $y^n - 1 = 0$ is related to circle division; if n is odd, he noted, its n roots are 1 and all the values of $\cos \frac{m}{n} 2\pi \pm \sqrt{-1} \sin \frac{m}{n} 2\pi$, for m an integer up to $\frac{n-1}{2}$; if n is odd we must also include -1 and otherwise use the same formula with m up to $\frac{n-2}{2}$. Multiplying $\sqrt[n]{\frac{a}{b}}$ by each of these in turn gives n values of y and therefore of x. Bezout showed that when $n = 3$ this procedure yields the three roots

$$x = \sqrt[3]{a^2 b} + \sqrt[3]{ab^2},$$
$$x = \sqrt[3]{a^2 b}\left(\tfrac{-1-\sqrt{-3}}{2}\right) + \sqrt[3]{ab^2}\left(\tfrac{-1+\sqrt{-3}}{2}\right),$$
$$x = \sqrt[3]{a^2 b}\left(\tfrac{-1-\sqrt{-3}}{2}\right) + \sqrt[3]{ab^2}\left(\tfrac{-1+\sqrt{-3}}{2}\right),$$

and when $n = 4$ the four roots

$$x = +\sqrt[4]{a^3 b} + \sqrt[4]{a^2 b^2} + \sqrt[4]{ab^3},$$
$$x = -\sqrt{-1}\sqrt[4]{a^3 b} - \sqrt[4]{a^2 b^2} + \sqrt{-1}\sqrt[4]{ab^3},$$
$$x = +\sqrt{-1}\sqrt[4]{a^3 b} - \sqrt[4]{a^2 b^2} - \sqrt{-1}\sqrt[4]{ab^3},$$
$$x = -\sqrt[4]{a^3 b} + \sqrt[4]{a^2 b^2} - \sqrt[4]{ab^3}.$$

Bezout also showed that another way of writing (14) is

$$x = (y^{n-1} + y^{n-2} + y^{n-3} + \cdots + y) \times b.$$

For him this was simply a convenient formula that avoided the need for the division he had done at (15), but it also gives immediately the expressions for x that he had just demonstrated for $n = 3$ and $n = 4$.

Finally, Bezout observed that division of a circle into equal parts was possible by ruler and compass construction alone when the number of sides belongs to one of the progressions 2, 4, 8, 16, … or 3, 6, 12, 24, … or 5, 10, 20, 40, … or 15, 30,

[12] Simpson 1750, II, 352–354.

60, 120, …. For equations of those degrees, he claimed, one might be able to express the roots in terms of radicals; but in all other cases only in terms of sines and cosines.

Bezout concluded his paper by posing a 'Problem III' (§23), which he promised to address in further work, namely, to determine precisely which classes of equation are solvable in terms of two, three, four, or more radicals of the same degree as the equation. In January 1763, he presented some preliminary findings to the Paris Academy, but they were not published until 1769, by which time Euler's 'De resolutione aequationum cuisvis gradus' had also been published. In this second paper Bezout did indeed examine several special classes of equations that can be solved algebraically and whose solutions are sums of two, three, or even four or five radicals, and his lists included all the equations identified by de Moivre in 1707, by Euler in 1738, and many others. That work will be examined in greater detail in Chapter 8.

For now, what matters far more than Bezout's lists of special cases is the method he proposed in 1762 for finding them, by transforming a given equation into another of the form $y^n - h = 0$. Some eighty years earlier Tschirnhaus too had suggested that it should be possible to find substitutions that would remove all the intermediate terms of an equation of degree n but there is no evidence that Bezout was aware of Tschirnhaus's work. The transformation that Bezout tried out in 1762 $\left(y = \frac{x+a}{x+b}\right)$ was in fact more restricted than the one that Tschirnhaus had proposed in 1683 ($y = x^{n-1} + ax^{n-2} + \cdots + h$) because it works only where the roots of the original equation lie on a circle. In any case Bezout's motivation came not from Tschirnhaus but directly from Euler's paper of 1738, and thus indirectly from de Moivre's of 1707.

Summary

De Moivre in 1707 was the first writer to identify a general class of equations of degree higher than four that could be solved algebraically: angle division equations of odd degree. Some preliminary steps in this direction had been taken by Dulaurens back in 1667 for equations up to degree 11, but Newton's multiple angle formulae enabled de Moivre to generalize the process to any odd degree. In all cases the solution consisted of a sum of two radicals of degree n, with square roots nested inside them.

It is likely that Euler first became aware of de Moivre's paper when it was republished by 'sGravesande in 1732. The following year, in a leap of faith, he conjectured that the solution of any equation of degree n (without a term of degree $n-1$) might be expressible as a sum of $n-1$ radicals of degree n. His evidence was slender, however, consisting only of equations of degree 2, 3, and 4, and the special cases identified by de Moivre. Euler developed his ideas considerably further in the 1750s, and was able to suggest a formula that might give not just one root but all n roots of a given equation. He managed to identify special classes of equation of degree five for which his hypothesis held but was very far from being able to prove it generally.

Meanwhile, Bezout too had taken up Euler's conjecture about roots as sums of radicals. His approach was to propose a transformation which in certain cases would convert a given equation into a circle division equation. By this means Bezout, like

Euler, was able to identify special classes of equations of degree n whose roots were sums of two, three, four, or five radicals of degree n. Further, he noted that the resolvent or reduced equation in such cases was at worst quadratic.

Both Euler and Bezout used the idea of roots as sums of radicals primarily to identify classes of equations that could be solved algebraically, and to both of them this appears to have been an important thing to do.[13] In some ways their position was exactly equivalent to Cardano's in the mid-sixteenth century: faced with the difficulty of solving equations in general, it nevertheless remained possible and indeed potentially useful to identify special cases that would yield to special methods.

[13] See also Euler 1790.

Chapter 6
Functions of the roots

Euler's work on topics that can be broadly classified as the theory of equations comprises no more than a tiny fraction of his total output: about twenty publications out of almost nine hundred catalogued by Gustav Eneström in the early twentieth century.[1] In contrast to his output on other subjects, Euler never gave the theory of equations consistent or prolonged attention. Before 1770 his writings on the subject were particularly sparse, no more than about seven papers in all, produced at irregular and widely spaced intervals. Nevertheless, almost every one of those papers contained a result or insight that others were prepared to pursue even if Euler himself did not. We have seen in Chapter 5 that his conjecture of 1733, that the roots of equations of degree n might be expressible as sums of up to $n-1$ radicals of degree n led to fruitful explorations by Bezout. Some thirteen years later, in 1746, Euler threw out another and quite different idea which was eventually to prove equally powerful: that one might investigate properties not just of the roots themselves but of *functions* of the roots. Euler did not at the time seem to regard this as a particularly significant suggestion: his initial presentation of it was just a short section of a long paper with a quite different objective. As with his conjecture of 1733 it was Bezout who recognized the implications; indeed Bezout was led to a hypothesis that directly contradicted Euler's views on the likely degree of resolvent equations. This chapter will examine (i) the context of Euler's work on a particular function of the roots, in 1746, (ii) Bezout's disagreement with Euler, and (iii) some of the work of Maclaurin and Euler on some special functions of the roots, the symmetric functions.

Euler and the factorization of polynomials

Euler left St Petersburg for Berlin in 1741 at the invitation of King Frederick II of Prussia, who hoped to reorganize the Berlin–Brandenburg Society of Scientists into an Academy comparable with that in Paris. Euler became mathematical director of the new Academy in 1743 and presented papers regularly at the fortnightly meetings, to audiences of typically twelve to twenty participants. One of the problems he attempted in his early years in Berlin was to prove that a polynomial with real coefficients could always be decomposed into linear and quadratic factors with real coefficients, part of a theorem that later came to be called the Fundamental Theorem of Algebra.[2] Euler presented his proof to the Academy on 10 November 1746, under the title 'Recherches sur les racines imaginaires des equations' ('Researches on the imaginary roots of equa-

[1] For a classification by subject of Euler's writings see http://www.math.dartmouth.edu/~euler/

[2] The Fundamental Theorem holds also for polynomials with complex coefficients, but Euler worked with real coefficients only.

tions').[3] Publishing delays in Berlin were not quite so severe as in St Petersburg but nevertheless significant: although the paper was first presented in 1746 it was not included in the Academy's *Mémoires* until 1749, printed in 1751. The published version of the paper is more than 60 pages long, written in a clear and very leisurely style.

Euler's motivation for this paper came not from an interest in solving equations but from the practicalities of the integral calculus. In fact he regarded imaginary roots as being of little use as solutions of equations but argued that they were of considerable help in analysis in general, in particular in the integration of algebraic fractions (§5). In such cases one needs to find the linear factors, real or imaginary, of the denominator; reciprocals of imaginary factors integrate to imaginary logarithms, which by appropriate substitutions can be converted to real circular functions. It was commonly assumed that such factorization was possible but such was the importance of the assumption that Euler wanted to provide a firm proof.[4]

Euler observed early in the paper (§6, §7) that imaginary roots occur in conjugate pairs to produce factors of the form $xx + px + q$, where p and q are real and q is necessarily positive (by which Euler meant strictly positive). If one accepts the truth of the theorem that an equation of degree n has n real or imaginary roots, then a simple lemma follows immediately, namely, that an equation of even degree with a negative final term must have at least two real roots. That theorem itself, however, was precisely what Euler was trying to prove. He therefore offered an alternative proof of the lemma based on reasoning from the curve of the polynomial $y = x^{2m} + Ax^{2m-1} + \cdots - OO$, that is, with a negative final term (by which Euler meant strictly negative) (§25). He had argued earlier (§22) that when $x = +\infty$ then also $y = +\infty$; and also when $x = -\infty$ then $y = (-\infty)^{2m} = +\infty$. That argument hardly meets modern standards of rigour but one can agree with Euler's conclusion that for large enough positive or negative values of x (what he called *positives infinies* or *negatives infinies*) the curve will lie above the x-axis. But when $x = 0$ we have $y = -OO$, which is negative. It is clear, therefore, that the curve must cross the x-axis at least twice, and each intersection corresponds to a real root, one positive and one negative.

The next part of Euler's paper consists of a lengthy inductive argument in which he aimed to show (though ultimately unsuccessfully) that every equation of even degree has real quadratic factors. The part of his argument that most concerns us here is his argument for equations of degree four (§27). In the usual way we may take a general quartic to be of the form

$$x^4 + Bx^2 + Cx + D = 0. \tag{1}$$

Euler wanted to prove that this can always be factorized as

$$(xx + ux + \alpha)(xx - ux + \beta) = 0 \tag{2}$$

[3]The presentation was recorded in the Academy *Registre* for 1746–66, BBAW MS I–IV–31, f. 8v. The paper is now known as E170.

[4]Euler's *Introductio ad analysin infinitorum* written during the 1740s contains a significant amount of material on decomposition into partial fractions.

with α, β, and u real. Comparing coefficients in (1) and (2) he arrived at the following equations for u, α, and β:

$$2\alpha = uu + B - \frac{C}{u}, \tag{3}$$

$$2\beta = uu + B + \frac{C}{u}, \tag{4}$$

and

$$u^6 + 2Bu^4 + (BB - 4D)uu - CC = 0. \tag{5}$$

The last was the equation Descartes had arrived at in 1637 (see page 47). Unlike Descartes, however, Euler was not interested in solving equation (5), only in showing that it has real solutions. His argument followed immediately from his lemma. The final term of (5) is negative and therefore (5) has at least two real roots. Substituting either of these real values into (3) and (4) we obtain values of α and β that are also real.[5] Thus we can always factorize (1) into real quadratic factors, as required.

Euler's purpose was to proceed by analogy to equations of higher degree, but he recognized that in such cases an explicit equation for u would be much harder to find. He therefore wished to show by 'reasoning alone' (*le seul raisonnement*) that the equation for u must be of even degree with its final term negative.[6] This was his argument. Suppose the four roots of (1) are $\mathfrak{a}$, $\mathfrak{b}$, $\mathfrak{c}$, $\mathfrak{d}$. Because the equation has no term in x^3 we know that

$$\mathfrak{a} + \mathfrak{b} + \mathfrak{c} + \mathfrak{d} = 0.$$

Further, we see from (2) that u is the sum of just two of these roots and $-u$ is the sum of the other two. There are thus just $\frac{4\times 3}{2\times 1} = 6$ possibilities for u, namely

$$\begin{aligned} u &= \mathfrak{a} + \mathfrak{b}, \quad & u &= \mathfrak{a} + \mathfrak{c}, \quad & u &= \mathfrak{a} + \mathfrak{d}, \\ u &= \mathfrak{c} + \mathfrak{d}, \quad & u &= \mathfrak{b} + \mathfrak{d}, \quad & u &= \mathfrak{b} + \mathfrak{c}. \end{aligned}$$

If we write $u = p$, $u = q$, $u = r$ for the three possibilities in the first row, then those in the second row are $u = -p$, $u = -q$, $u = -r$. Thus the equation for u is

$$(u - p)(u - q)(u - r)(u + p)(u + q)(u + r) = 0,$$

or

$$(uu - pp)(uu - qq)(uu - rr) = 0. \tag{6}$$

This is an equation of degree 6 in which only even powers of u appear, just as in (5). We can also see that the final term of (6) is $-ppqqrr$. Euler checked that this is always

[5] Euler ignored the possibility that $u = 0$ which, as can be seen from equation (5), arises only when $C = 0$, in which case equation (1) reduces to a quadratic in x^2.

[6] *L'une & l'autre de ces deux circonstances se peut découvrir par le seul raisonnement, sans qu'on ait besoin de chercher l'équation même qui renferme l'inconnue u.* [Both of these two conditions may be found by reasoning alone without any need to seek the actual equation containing the unknown u.] Euler (1749) [1751], §33.

L'une & l'autre de ces deux circonſtances ſe peut découvrir par le ſeul raiſonnement, ſans qu'on ait beſoin de chercher l'équation même, qui renferme l'inconnue u. Donc puisque dans la ſuite, où je paſſerai à des équations de plus hauts degrés, il ſeroit trop difficile & même impoſſible de trouver l'équation, par laquelle l'inconnue u eſt déterminée; il ſera important de découvrir les deux circonſtances mentionnées par le ſeul raiſonnement, pour l'équation propoſée du quatriéme degré, afin de frayer le chemin pour mettre en uſage ce même raiſonnement, lorsque l'équation propoſée ſera d'un plus haut degré.

Soit donc l'équation propoſée dégagée déjà du ſecond terme

$$x^4 + Bx^2 + Cx + D = 0$$

& poſant les quatre racines de cette équations

$$x = \mathfrak{a}; \quad x = \mathfrak{b}; \quad x = \mathfrak{c}; \quad x = \mathfrak{d}$$

il eſt d'abord clair que la ſomme de ces quatre racines

$\mathfrak{a} + \mathfrak{b} + \mathfrak{c} + \mathfrak{d}$ ſera égale à zero.

Enſuite poſant en général un des facteurs doubles de cette équation $xx - ux + \beta$ ou $xx - ux + \beta = 0$, il eſt certain que u ſera la ſomme de deux racines quelconques des quatre ſuppoſées $\mathfrak{a}, \mathfrak{b}, \mathfrak{c}, \mathfrak{d}$. Donc cette lettre u regardée comme notre inconnüe peut avoir autant de valeurs differentes, qu'il y a de diverſes combinaiſons de deux lettres priſes de ces quatre $\mathfrak{a}, \mathfrak{b}, \mathfrak{c}, \mathfrak{d}$. Or ce nombre de combinaiſons étant comme on ſait $= \frac{4 \cdot 3}{1 \cdot 2} = 6$, la lettre u eſt ſuſceptible de 6 valeurs differentes, & non de pluſieurs. Donc la lettre u ſera déterminée par une équation du 6me degré, qui aura les ſix racines ſuivantes:

$$\text{I. } u = \mathfrak{a}+\mathfrak{b}; \quad \text{II. } u = \mathfrak{a}+\mathfrak{c}; \quad \text{III. } u = \mathfrak{a}+\mathfrak{d}$$
$$\text{IV. } u = \mathfrak{c}+\mathfrak{d}; \quad \text{V. } u = \mathfrak{b}+\mathfrak{d}; \quad \text{VI. } u = \mathfrak{b}+\mathfrak{c}$$

Donc puisque $\mathfrak{a}+\mathfrak{b}+\mathfrak{c}+\mathfrak{d} = 0$, ſi nous poſons les trois premiéres de ces ſix racines:

$$\text{I. } u = p; \quad \text{II. } u = q; \quad \text{III. } u = r$$

les

Roots of a quartic taken in pairs, from Euler (1746).

negative even when two or four of $\mathfrak{a}$, $\mathfrak{b}$, $\mathfrak{c}$, $\mathfrak{d}$ are imaginary. We can therefore be certain that (6) always has at least two real roots.

Euler immediately tried to extend this argument to equations of degree eight (§34). Suppose such an equation is

$$x^8 + Bx^6 + Cx^5 + Dx^4 + Ex^3 + Fx^2 + Gx + H = 0,$$

which Euler hoped to factorize into two real quartics as

$$(x^4 + ux^3 + \alpha x^2 + \beta x + \gamma)(x^4 - ux^3 + \delta x^2 + \epsilon x + \zeta) = 0.$$

Euler claimed that by equating coefficients it is possible to eliminate α, β, γ, δ, ϵ, ζ 'without extraction of roots' (*sans besoin d'aucune extraction de racine*),[7] and so to arrive at an equation for u. Since u here represents the sum of any four roots, the equation must be of degree $\frac{8.7.6.5}{4.3.2.1} = 70$. If p is a possible value of u then so is $-p$ and so, Euler argued, the equation must contain 35 factors of the form $(uu - pp)$. Thus its final term must be negative and so it must have at least two real roots. Unfortunately, there was a fatal flaw in his argument: to determine values of $\alpha, \beta, \gamma, \delta, \epsilon, \zeta$ in terms of $u, B, C, D, \ldots$ root extraction *is* required. There is therefore no guarantee that a real value of u will lead to real values of α, β, γ, δ, ϵ, ζ. Nevertheless, Euler continued his argument inductively to all equations of degree 2^n and then to those of degree $2^n + 2m$, where n, m are integers.

For our purposes, the most important feature of Euler's paper is not his laboured and ultimately incorrect pursuit of factors, but his construction of equation (6), an equation whose roots are sums of pairs of roots of (1). Euler did not seem to see any great significance in this. Bezout, however, certainly did.

At the beginning of his 'Mémoire sur plusieurs classes d'équations [...] qui admettent une solution algébrique' (1762) [1764], Bezout made some important observations on the degree of resolvent equations. In the method of Descartes for quartics, he argued, it was easy to see that the resolvent must be of degree 6 because each of its roots is a sum of two roots of the original equation, and such a sum can take six possible values. This was precisely what Euler had shown in detail in his 'Recherches sur les racines imaginaires' but Bezout did not mention that paper and probably arrived at the same conclusion independently. Bezout also recognized the consequences. Suppose we try to solve an equation of degree 5 in a similar way, he argued, that is, by factorizing it into a quadratic and a cubic. There are ten possible ways of forming a sum of two (or three) of the original roots, so the resolvent equation will be of degree 10 (as Hudde had also discovered, but Bezout did not refer to him either). Further, the fifth roots that Bezout by now expected to find in the solution to the original (see Chapter 5) were not going to arise from the quadratic or cubic factors but only from the equation of degree 10

[7] *De ces égalites on eliminera successivement les lettres α, β, γ, δ, ϵ, ζ, ce qui se pourra faire, comme on sait, sans qu'on ait besoin d'aucune extraction de racine;* [From these equations one may eliminate successively the letters α, β, γ, δ, ϵ, ζ, which may be done, as one knows, without any need for extraction of roots;] Euler (1749) [1751], §34.

itself. That is, the equation of degree 10 must pose at least the same difficulties as the original equation of degree 5.

Although Bezout did not say as much, these observations contradicted Euler, who in 'De formis radicum aequationum' (1733) [1738] had made a different conjecture. There Euler had shown that a quadratic equation has a resolvent of degree 1, a cubic has a resolvent of degree 2, and a quartic has a resolvent of degree 3 (see pages 109–111). He had therefore supposed that an equation of degree n will in general have a resolvent of degree $n-1$. This was an optimistic hypothesis, because it meant that the solution of equations of any degree could open up the solution of equations one degree higher. Bezout's insight, on the other hand, was profoundly pessimistic, because it suggested that the resolvent was going to be, in general, at least as problematic as the original equation. The resolution of this issue will be discussed in Chapter 8.

Symmetric functions of the roots

The composition of the coefficients of an equation in terms of its roots had been clear since the publication of Harriot's *Praxis* (1631). In eighteenth-century notation, if an equation

$$x^n - Ax^{n-1} + Bx^{n-2} - Cx^{n-2} + \cdots \pm M = 0$$

has n roots $a, b, c, d, \ldots$, then

$$\begin{aligned} A &= a + b + c + \cdots, \\ B &= ab + ac + ad + \cdots, \\ C &= abc + abd + bcd + \cdots, \\ &\cdots. \end{aligned}$$

The coefficients $A, B, C, \ldots, M$ are known as *symmetric functions* of the roots because they do not change if the roots are permuted amongst themselves. It follows that a given set of roots will always give rise to the same unique equation.

Girard in his *Invention nouvelle* (1629) recognized that there are other functions of the roots that also remain invariant when the roots are permuted, for example, the sum of squares, sum of cubes, and so on. Girard's equations connecting such sums to the coefficients are given on page 46 above. Newton in his *Arithmetica universalis* (1707) wrote a similar set of equations but in recursive form. If P is the sum of the roots, Q the sum of their squares, R the sum of their cubes, and so on, Newton's equations are[8]

$$\begin{aligned} P &= A, \\ Q &= AP - 2B, \\ R &= AQ - BP + 3C, \\ S &= AR - BQ + CP - 4D, \\ &\cdots. \end{aligned}$$

[8]Newton's original equations look slightly different from these because he used $p, q, r, \ldots$ for the coefficients (all with negative sign) and $a, b, c, \ldots$ for sums of powers (see page 73 above). Newton 1707, 251–252.

As so often, Newton offered a few numerical examples to illustrate his rules, but no general proof of their validity. In the remainder of this section we will examine proofs provided almost half a century later, first by Maclaurin and then by Euler.

Maclaurin, who elucidated so much of Newton's *Arithmetica universalis*, included a proof of Newton's formulae in a letter to Philip Stanhope, a Fellow of the Royal Society, in July 1743. It was later incorporated into his *Treatise of algebra*.[9] For an equation $x^n - Ax^{n-1} + Bx^{n-2} - Cx^{n-2} + \cdots - Lx + M = 0$ with roots $a, b, c, \ldots$ and any $r \geq n$ Maclaurin could write[10]

$$\begin{aligned}
a^r - Aa^{r-1} + Ba^{r-2} - Ca^{r-3} + \cdots - La^{r-n+1} + Ma^{r-n} &= 0,\\
b^r - Ab^{r-1} + Bb^{r-2} - Cb^{r-3} + \cdots - Lb^{r-n+1} + Mb^{r-n} &= 0,\\
c^r - Ac^{r-1} + Bc^{r-2} - Cc^{r-3} + \cdots - Lc^{r-n+1} + Mc^{r-n} &= 0,\\
&\ldots.
\end{aligned}$$

For convenience we will introduce notation that Maclaurin did not use, namely, $S_r = a^r + b^r + c^r + \cdots$. Adding the above equations immediately gives the general rule for $r \geq n$,

$$S_r = AS_{r-1} - BS_{r-2} + CS_{r-3} - \cdots - MS_{r-n}.$$

When $r < n$ the matter is more difficult. Maclaurin handled such cases one by one, beginning with $r = n - 1$, where he could write

$$\begin{aligned}
a^{n-1} - Aa^{n-2} + Ba^{n-3} + \cdots - L + \frac{M}{a} &= 0,\\
b^{n-1} - Ab^{n-2} + Bb^{n-3} + \cdots - L + \frac{M}{b} &= 0,\\
c^{n-1} - Ac^{n-2} + Bc^{n-3} + \cdots - L + \frac{M}{c} &= 0,\\
&\ldots.
\end{aligned}$$

From the composition of the coefficients Maclaurin knew that

$$L = \frac{M}{a} + \frac{M}{b} + \frac{M}{c} + \cdots,$$

and so adding the equations as before he had

$$S_{n-1} = AS_{n-2} - BS_{n-3} + CS_{n-4} - \cdots + (n-1)L.$$

The case $r = n - 2$ can be handled similarly but the calculations are longer (taking up two pages in Maclaurin's *Treatise of algebra*).[11] For the general case Maclaurin fell back on some of the formulae he had derived in his work on the number of impossible

[9] Maclaurin 1748, 285–295.

[10] Maclaurin's notation suggests that the equation is of even degree but his argument does not depend on such an assumption.

[11] Maclaurin 1748, 288–289.

roots (see Chapter 4) so his proof was no longer self-contained, and it becomes less and less transparent as it proceeds.

Before Maclaurin's proof appeared in print, Euler also turned his attention to the problem, having come across it, presumably, in the *Arithmetica universalis*. His paper entitled 'Demonstratio gemina theorematis Neutoniani' ('A double demonstration of the Newtonian theorem') was presented to the Berlin Academy on 12 January 1747, just two months after his 'Recherches sur les racines'.[12] The 'Demonstratio' was not published in the *Mémoires* of the Academy, however, but in a collection of short papers entitled *Opuscula varii argumenti* (*Short works on various matters*), published in 1750.

Euler offered two proofs of Newton's rules, one which made use of calculus and infinite series, the other purely algebraic. For his first proof (§5) Euler supposed that the equation

$$x^n - Ax^{n-1} + Bx^{n-2} - Cx^{n-3} + \cdots \mp Mx \pm N = 0 \tag{7}$$

has roots α, β, $\ldots\nu$, so that he could write

$$Z = x^n - Ax^{n-1} + Bx^{n-2} - Cx^{n-3} + \cdots \pm N = (x-\alpha)(x-\beta)\ldots(x-\nu).$$

Taking logarithms and differentiating gave him

$$\frac{dZ}{Zdx} = \frac{1}{x-\alpha} + \frac{1}{x-\beta} + \cdots + \frac{1}{x-\nu}.$$

Further, each term on the right could in turn be written as an infinite series, for example

$$\frac{1}{x-\alpha} = \frac{1}{x} + \frac{\alpha}{x^2} + \frac{\alpha^2}{x^3} + \cdots.$$

For sums of powers of the roots Euler used the notation

$$S\alpha^p = \alpha^p + \beta^p + \cdots + \nu^p,$$

so he now had

$$\frac{dZ}{Zdx} = \frac{n}{x} + \frac{1}{x^2}S\alpha + \frac{1}{x^3}S\alpha^2 + \cdots. \tag{8}$$

But differentiating Z directly he could also write

$$\frac{dZ}{Zdx} = \frac{nx^{n-1} - (n-1)Ax^{n-2} + \cdots \mp M}{x^n - Ax^{n-1} + Bx^{n-2} - \cdots \pm N}. \tag{9}$$

[12]The presentation was recorded in the Academy *Registre* for 1746–66, BBAW MS I–IV–31, f. 11. A manuscript copy of the paper, now known as E153, is held by the Academy as BBAW MS I–M 80, C.5, 7–11.

Multiplying both (8) and (9) by the denominator of (9) and then equating coefficients he arrived correctly at Newton's formulae

$$\begin{aligned} S\alpha &= A, \\ S\alpha^2 &= AS\alpha - 2B, \\ S\alpha^3 &= AS\alpha^2 - BS\alpha + 3C, \\ S\alpha^4 &= AS\alpha^3 - BS\alpha^2 + CS\alpha - 4D, \end{aligned}$$

and so on.

Euler's second demonstration (§8), was algebraic. For $S\alpha^r$ where $r \geq n$, his proof was exactly the same as Maclaurin's. Thus, for $m \geq 0$ he had

$$S\alpha^{n+m} = AS\alpha^{n+m-1} - BS\alpha^{n+m-2} + \cdots \mp MS\alpha^{m+1} \pm NS\alpha^m. \tag{10}$$

His argument for sums of powers for $m < 0$, however, was rather more subtle than Maclaurin's. As we saw in Maclaurin's treatment, awkward fractions appear in (10) when m takes negative values. Euler avoided these by constructing a new sequence of equations:

$$\begin{aligned} x - A &= 0, \\ x^2 - Ax + B &= 0, \\ x^3 - Ax^2 + Bx - C &= 0, \\ x^4 - Ax^3 + Bx^2 - Cx + D &= 0. \end{aligned}$$

Each of these, he argued, will have its own roots, different from those for the other equations in the list or for (7), but in each case the sum of the roots will be A and the sum of their products in pairs will be B. Thus the sums of squares of the roots, which depend only on A and B, will be expressed by the same rule for each of the above equations and also for (7). Similarly the sums of cubes of the roots can always be expressed in the same way in terms of A, B, and C; and so on for sums of higher powers. Applying condition (10) with $m = 0$ to each equation in turn, therefore, Euler claimed that

$$\begin{aligned} S\alpha &= A, \\ S\alpha^2 &= AS\alpha - 2B, \\ S\alpha^3 &= AS\alpha^2 - BS\alpha + 3C, \end{aligned}$$

and so on, just as Newton had claimed.

We may mention briefly here that Euler returned to formulae for sums of powers of the roots of an equation in 1770, in a paper entitled 'Observationes circa radices aequationum' ('Observations on roots of equations'). Starting from Newton's recursive formulae for the sums of nth powers of the roots, Euler expressed those same formulae in closed form as infinite series:

$$Sx^n = A^n + PA^{n-2} + QA^{n-3} + RA^{n-4} + \cdots \tag{11}$$

where P, Q, R, ... are polynomials in B, C, D, ..., H, ... Euler's calculation of the coefficient of A^{n-8}, for instance, gave him

$$nH + \tfrac{n(n-7)}{1.2}(2BF + 2CE + DD) + \tfrac{n(n-6)(n-7)}{1.2.3}(3B^2D + 3BC^2) + \tfrac{n(n-5)(n-6)(n-7)}{1.2.3.4}B^4.$$

Newton's finite formulae can be recovered from Euler's infinite series by neglecting terms containing negative powers of A. Euler, however, began to investigate the meaning of (11) when all its terms are retained. He came to the remarkable conclusion that although (11) in its truncated form expresses the sum of nth powers of the roots of an equation, in its infinite form it gives the nth power of the largest root.[13] Euler developed his series for powers of the roots further during the remaining years of his life but that work takes us beyond the scope of the present investigation.[14]

Summary

We have seen in this chapter and the previous one that in 1733 and 1746 Euler came to two important new insights concerning polynomial equations. The first was his conjecture that roots of such equations could be expressed as sums of radicals. The second was the possibility of constructing new equations whose roots were functions of the roots of a given equation. Euler himself, perhaps because he was creatively engaged in so many different areas of mathematics, failed at first to pursue either idea very far, and by the time he began to pay more serious attention to equation-solving in the early 1760s Bezout had caught up with him, and was beginning to understand the implications of those ideas more clearly than Euler himself had done. Indeed Bezout was led to question Euler's assertion that an equation of degree n would always have a resolvent equation of degree $n-1$. We will return to that discussion in Chapter 8. Before that, however, we need to explore another strand of investigation that was becoming increasingly important, the theory of elimination.

[13] Euler (1770) [1771], §VII.

[14] See Euler (1779) [1783], 1789a, 1789b, 1801.

Chapter 7
Elimination theory

In this chapter we follow a further eighteenth-century development in the understanding of equations, namely, the theory of elimination. The first hints of it had appeared in Newton's *Arithmetica universalis* in 1707 but, as with the idea of roots as sums of radicals (Chapter 5), and of equations in functions of the roots (Chapter 6), it was Euler who wrote the paper that began to establish the theory properly. For him the subject arose naturally out of his work on curves in 1747 and 1748. Almost simultaneously the theory was also pursued and developed by Gabriel Cramer. Euler took it up again in the early 1750s though this later work was not published until 1764. By that time Bezout, whose thoughts so often seemed to run parallel to Euler's, had also taken up the subject of elimination. It was Lagrange, however, who gave the clearest and most general exposition, in 1769. Lagrange had contributed little or nothing to the theory of equations until then but from that point onwards was to become the leading figure in the story. We will begin, however, with the simple but thought-provoking results offered by Newton in 1707.

Newton's elimination of quantities, 1707

Newton's instructions in the *Arithmetica universalis* for handling equations included a section entitled 'De duabus pluribus aequationibus in unam transformandis ut incognitae quantitates exterminentur' ('On transforming two or more equations into one, in order to eliminate unknown quantities').[1] His first method was to find (if possible) explicit expressions for an unknown from each of two equations and then equate them. As an example, he gave the equations

$$\begin{aligned} ax - 2by &= ab, \\ xy &= bb, \end{aligned}$$

from which he derived (without worrying about whether $x = 0$)

$$\begin{aligned} y &= \frac{ax - ab}{2b}, \\ y &= \frac{bb}{x}. \end{aligned}$$

Equating these gives a quadratic equation for x, that is, an equation of higher degree in x than either of the originals,

$$axx - abx - 2b^3 = 0. \tag{1}$$

[1] Newton 1707, 69–76; 1720, 60–67.

When it is less easy, or indeed impossible, to find explicit expressions for the unknowns, an alternative method is substitution. As an example Newton gave the equations

$$ayy + aay = z^3,$$
$$yz - ay = az.$$

Here Newton substituted $y = \dfrac{az}{z-a}$ from the second equation into the first (without worrying whether $z = a$) to give, after clearing fractions,

$$z^4 - 2az^3 + aazz - 2a^3z + a^4 = 0. \tag{2}$$

As was the case for x in (1), z appears in (2) with higher degree than in either of the original equations.

In both the above examples one of the original equations was linear. When both equations are of degree 2 or higher the problem becomes harder. Suppose we have, as in another of Newton's examples, the simultaneous equations

$$xx + 5x = 3yy, \tag{3}$$
$$2xy - 3xx = 4. \tag{4}$$

Equating expressions for $3xx$ obtained from each of these gives

$$9yy - 15x = 2xy - 4,$$

which is linear in x. From this Newton was able to substitute

$$x = \frac{9yy + 4}{2y + 15}$$

into (3) to arrive at

$$69y^4 - 90y^3 + 72yy + 40y + 316 = 0. \tag{5}$$

As before, the final equation, the 'elimination equation' (5), is of higher degree than either of the original equations.

Newton himself made no observations about the degree of the final equation but remarked only that the process of elimination could be extremely laborious (*maxime laboriosus*). As an aid to such calculations, therefore, he offered four 'rules', of which Rule I is the following. If x is to be eliminated from $axx + bx + c = 0$ and $fxx + gx + h = 0$ it must be the case that

$$\overline{ah - bg - 2cf} \times ah + \overline{bh - cg} \times bf + \overline{agg + cff} \times c = 0. \tag{6}$$

Thus, for instance, take the two equations (3) and (4) above, now written as

$$xx + 5x - 3yy = 0,$$
$$3xx - 2xy + 4 = 0.$$

Newton's Rule I tells us that it is possible to eliminate x from these two equations only if

$$(4 + 10y + 18yy) \times 4 + (20 - by^3) \times 15 + (4yy - 27yy) \times -3yy = 0,$$

that is, if

$$316 + 40y + 72yy + 300 - 90y^3 + 69y^4 = 0,$$

as he already found by other means at (5).

In Rules II and III Newton replaced the first equation, $axx + bx + c = 0$, by cubic or quartic equations; in Rule IV both equations were cubic. As one might expect, the conditions that the coefficients must satisfy become increasingly complicated, and contain terms of degree higher than those in either of the original equations. Newton's example of a quadratic and cubic equation, for instance, leads to an elimination equation of degree 6.

Euler's first paper on elimination, 1748

On 12 October 1747 Euler presented to the Berlin Academy a paper in which he considered the number of points needed to fix curves of order 3, 4, or 5 respectively.[2] On 18 January 1748 he followed it up with a second paper that clearly stemmed from the same research, 'Demonstration sur le nombre des points, ou deux lignes des ordres quelconques peuvent se couper' ('Demonstration of the number of points in which two curves of any order may intersect').[3] The two papers were published side by side in the *Mémoires* of the Academy in for 1748, printed in 1750. In the second paper, the 'Demonstration', Euler set out to prove that two curves of order m and n, respectively, can intersect in up to mn points.[4] The truth of this proposition, he remarked, was already accepted by geometers on the evidence of many particular cases, but he wished to give a rigorous and general demonstration of it.

Euler began, as he so often did, by building upwards from easy examples. Where $m = 1$, for example, that is, where one of the curves is a straight line, it is easy to show that it must intersect a curve of order n up to n times (§4). If $m = 2$, and the curve is a parabola with equation $y = axx + bx + c$, then it is similarly easy to show that it intersects a curve of order n up to $2n$ times (§7).

These are simple cases, however, and in general it is much more difficult to see what the degree of the elimination equation should be. Indeed, Euler observed, one

[2]The presentation is recorded in the Academy *Registre* for 1746–66, BBAW MS I–IV–31, f. 21v. A manuscript copy of the paper, now known as E147, is held by the Academy as BBAW MS I–M 88, C.5, 228–235.

[3]The presentation is recorded in the Academy *Registre* for 1746–66, BBAW MS I–IV–31, f. 26. A manuscript copy of the paper, now known as E148, is held by the Academy as BBAW MS I–M 92, C.6, 14–21.

[4]Some of this work also appears in Euler 1748, II, §474–§485.

all too often arrives at an equation whose degree is higher than it need be.[5] Take, for example, the two cubic equations (§11)

$$Py^3 + Qy^2 + Ry + S = 0, \tag{7}$$

$$py^3 + qy^2 + ry + s = 0, \tag{8}$$

where P, Q, R, S, p, q, r, s are any functions of x. Euler argued that we may eliminate y^3 either by subtracting $S\times(8)$ from $s\times(7)$ (and dividing by y) to give

$$(Ps - pS)y^2 + (Qs - qS)y + (Rs - rS) = 0 \tag{9}$$

or by subtracting $P\times(8)$ from $p\times(7)$ to give

$$(Qp - qP)y^2 + (Rp - rP)y + (Sp - sP) = 0. \tag{10}$$

Note that (9) and (10) are both linear in $(Ps - pS)$. In exactly the same way we can eliminate y^2 from (9) and (10) in two different ways to give

$$\begin{aligned}((Ps - pS)(Sp - sP) - (Qp - qP)(Rs - rS))y\\ + (Qs - qS)(Sp - sP) - (Rp - rP)(Rs - rS) = 0,\end{aligned} \tag{11}$$

$$\begin{aligned}((Qs - qS)(Qp - qP) - (Rp - rP)(Ps - pS))y\\ - (Rs - rS)(Qp - qP) - (Sp - sP)(Ps - pS) = 0,\end{aligned} \tag{12}$$

where (11) and (12) are both quadratic in $(Ps - pS)$. Finally, eliminating y from (11) and (12) leads to an even longer equation (in the printed version of Euler's paper it spreads over four lines), this time of degree four in $(Ps - pS)$. Inspection reveals, however, that the entire equation can be divided through by $(Ps - pS)$, reducing it to degree three in $(Ps - pS)$.

If we take (7) and (8) to represent curves of order 3, then P and p must be constants, while Q and q are at most linear, R and r at most quadratic, and S and s at most cubic in x. Thus the final equation in x will be of degree at most nine. In other words, the two cubic curves represented by (7) and (8) will intersect in up to nine points, as one would expect.

The problem with this method is that superfluous factors introduced by repeated multiplication, like $(Ps - pS)$ above, may not always be easy to detect. Euler therefore proposed a different method of working in which one could be sure of arriving at an elimination equation of the correct degree (§16).

Suppose we wish to eliminate y from the two equations

$$y^m - Py^{m-1} + Qy^{m-2} - Ry^{m-3} + \cdots = 0, \tag{13}$$

$$y^n - py^{n-1} + qy^{n-2} - ry^{n-3} + \cdots = 0, \tag{14}$$

[5] *dans la plupart des cas si l'on se sert des methodes ordinaires d'eliminer on parviendra à une équation de plus de dimensions, que* mn*; de sorte qu'emploient cette maniere, on devroit plutot croire que la proposition fut fausse.* [in most cases, if one uses ordinary methods of elimination one even arrives at an equation of higher degree than mn; so that using such a method, one must usually think that the proposition is false.] Euler 1748a, §10.

where, as before, $P, Q, R, \ldots\ p, q, r, \ldots$ are functions of x. Euler argued that any function y of x (Euler called it a 'value' (*valeur*)) that satisfies one equation must also satisfy the other.[6] Suppose, then, that the solutions for y of (13) are the functions A, $B, C, \ldots$ (m in number) and of (14) are the functions $a, b, c, \ldots$ (n in number). Euler was making an enormous leap here from the theorem (which he assumed to be true) that an equation of degree n with numerical coefficients has n numerical roots. Now he was presuming that an equation of degree n in y whose coefficients are functions of x will similarly have n 'roots' which are themselves functions of x. Unfortunately, he made none of this explicit, but it enabled him to argue that any of $A, B, C, \ldots$ can be identified with any of $a, b, c, \ldots$ or, as Euler put it, each of the 'roots' of the first equation may be equal to each of the 'roots' of the second.[7] Thus the elimination equation must contain all possible factors

$$\begin{aligned}
&(A-a)(A-b)(A-c)\ldots,\\
&(B-a)(B-b)(B-c)\ldots,\\
&(C-a)(C-b)(C-c)\ldots,\\
&\ldots.
\end{aligned}$$

Setting this product equal to zero therefore gives the required equation.

Continuing to treat the functions $A, B, C, \ldots, a, b, c, \ldots$ as analogous to numerical roots, Euler now argued that from (14)

$$(y-a)(y-b)(y-c)\ldots = y^n - py^{n-1} + qy^{n-2} - ry^{n-3}\ldots.$$

The elimination equation is therefore the product of the m factors

$$\begin{aligned}
&A^n - pA^{n-1} + qA^{n-2} - rA^{n-3} + \cdots,\\
&B^n - pB^{n-1} + qB^{n-2} - rB^{n-3} + \cdots,\\
&C^n - pC^{n-1} + qC^{n-2} - rC^{n-3} + \cdots,\\
&\ldots.
\end{aligned}$$

Only now did Euler discuss 'expressions' for the 'roots' $A, B, C, \ldots$ observing that such expressions were often irrational, and indeed it may not be possible to find them explicitly.[8] Nevertheless, he claimed, we know that their sum is P, the sum of their products in pairs is Q, and so on. Further, he claimed, any expression in which the 'roots' $A, B, C, \ldots$ appear symmetrically (*également*) can be expressed in terms of $P, Q\ R, \ldots$. Euler made no attempt to prove this important claim, which he appears to have arrived at by pure intuition.[9]

[6] *Or d'abord on voit que la valeur de y, qui résulte d'une de ces équations doit être égale à la valeur de y, qui résulte de l'autre.* Euler 1748, §16.

[7] *il est clair que* […] *une des racines de la premier équation sera égale à une des racines de l'autre.* Euler 1748, §16.

[8] *Quoique les expressions des racines A, B, C, D, &c. & a, b, c, d, &c soient pour la pluspart irrationelles, & souvent telles, qu'on ne les peut assigner;* Euler 1748, §20.

[9] *Et par ces valeurs P, Q, R, S, &c. on est en état d'exprimer toutes les expressions, dans lesquelles entrent toutes les racines également, par des formules rationelles composées de P, Q, R, S, &c.* Euler 1748, §20.

Euler went on to say that if his exposition seemed obscure it was only because of its great generality, and all doubts would vanish once it was applied to a particular case. He therefore returned to the problem he had addressed earlier, that of finding the elimination equation for the two cubic equations

$$y^3 - Py^2 + Qy - R = 0,$$
$$y^3 - py^2 + qy - r = 0.$$

Assuming that the roots of these two equations are the functions A, B, C and a, b, c, respectively, Euler's theory enabled him to write down the elimination equation immediately as

$$(A^3 - pA^2 + qA - r)(B^3 - pB^2 + qB - r)(C^3 - pC^2 + qC - r) = 0.$$

After multiplication, this gave him an equation with 64 individual terms on the left hand side. Euler was able to rewrite it, however, using the properties $A+B+C = P$, $AB + BC + CA = Q$, and $ABC = R$, to arrive at an equation in P, Q, R, p, q, r containing only 34 terms, of which just the first six and the last three are given here

$$R^3 - pQR^2 + qQ^2R - 2qPR^2 - rQ^3 + 3rPQR - \cdots - p^3R^2 + 2qrQQ + ppqQR = 0.$$

Now assuming that P and p were linear, Q and q quadratic, and R and r cubic in x, Euler could check that, as he had predicted, every term of this final equation is of degree no higher than 9. That is, the elimination equation derived by this method is of the correct degree, with no superfluous factors.

Newton had given the same equation in his *Arithmetica universalis* for the elimination of an unknown from two cubics (Newton's Rule IV) but Euler did not mention it. He had read at least some of the *Arithmetica universalis* by 1746, because that year he proved Newton's rules for sums of powers of roots (see pages 128–129) but at the time may have overlooked the elimination rules.

Cramer's theory of curves, 1750

Gabriel Cramer was appointed professor of mathematics in Geneva in 1724 when he was twenty years old. To begin with he shared the work and the salary with another equally young mathematician, Giovanni Calendrini, under the unusual but rather imaginative condition that one of them would travel for two or three years while the other carried out full teaching duties in Geneva. Thus between 1727 and 1729 Cramer was able to work in Basel with Johann Bernoulli and for a short time Euler also. Later he met 'sGravesande in Leiden, Halley, de Moivre, and Stirling in London, and Fontenelle, Maupertuis, Clairaut, and others in Paris. As a result he must have been one of the most widely connected mathematicians of the period, and seems to have been universally respected by his many acquaintances and correspondents. He remained in post in Geneva, single-handedly after 1734, until his death in 1752.

Cramer's most important work was done during the 1740s, when he edited the collected works of both Johann and Jacob Bernoulli, and the correspondence between Johann Bernoulli and Leibniz. It was during this period that he also carried out his investigations into properties of curves. Cramer knew Newton's classification of cubic curves, his *Enumeratio linearum tertii ordinis* (1704), and also Stirling's detailed commentary on it, *Lineae tertii ordinis Neutonianae* (1717). Cramer extended similar methods of classification and analysis to curves of higher order and his findings were published in his *Introduction à l'analyse des lignes courbes* (*Introduction to the analysis of curved lines*) in 1750. Thus Cramer's *Introduction* and Euler's 'Demonstration' appeared in print in the same year, though both had been completed some time earlier and in Cramer's case had probably been in preparation for some years. It seems unlikely that Cramer and Euler knew the details of each other's work before publication; indeed, Cramer admitted in his Preface that Euler's *Introductio ad analysin infinitorum* (1748) would have been useful to him if only he had seen it in time.

In the third chapter of his *Introduction* Cramer claimed (as had Euler in 1748) that two curves of order m and n, respectively, will intersect in up to mn points.[10] For a proof he referred his reader to an Appendix. His argument there was similar in principle to Euler's in the 'Demonstration', but his style of presentation was quite different, as can be seen from the following outline of his treatment, given in his own notation.[11]

Suppose we have two equations

$$x^n - [1]x^{n-1} + [1^2]x^{n-2} - [1^3]x^{n-3} + \cdots + [1^n] = 0, \tag{A}$$

$$(0)x^0 + (1)x^1 + (2)x^2 + (3)x^3 + \cdots + (m)x^m = 0, \tag{B}$$

where $[1^r]$ represents a rational function in a second variable y, of degree no more than r, and the notation (s) represents a rational function in y of degree no more than $m-s$. Now suppose that $a, b, \ldots$ are the n roots of (A). As in Euler's treatment, these 'roots' are themselves functions of y. Cramer was more explicit on this point than Euler had been, claiming that they are rational or irrational functions of y which satisfy (A), and that since (A) is of degree n there must be n of them.[12] Substituting these functions into (B) we have n equations in y:

$$\begin{aligned}(0)a^0 + (1)a^1 + (2)a^2 + (3)a^3 + \cdots + (m)a^m &= 0, & (\alpha)\\ (0)b^0 + (1)b^1 + (2)b^2 + (3)b^3 + \cdots + (m)b^m &= 0, & (\beta)\\ (0)c^0 + (1)c^1 + (2)c^2 + (3)c^3 + \cdots + (m)c^m &= 0, & (\gamma)\\ &\cdots\end{aligned}$$

[10]Cramer 1750, 76.

[11]Cramer 1750, 660–676.

[12]*Que a, b, c, d, &c. représentent les racines de l'éq: A, ou les valeurs de x dans cette équation* $x^n - [1]x^{n-1} + \cdots + [1^n] = 0$. *Comme elle est du d'égré n, le nombre de ses racines est n.* [That a, b, c, d, etc. represent the roots of equation (A), or the values of x in the equation $x^n - [1]x^{n-1} + \cdots + [1^n] = 0$. Since it is of degree n the number of its roots is n.] Cramer 1750, 660.

660 APPENDICE.

No. II.

Voyez pag. 76.

§. 1. Soient x & y deux grandeurs variables, dont le rapport, tant entr'elles qu'avec des grandeurs constantes, soit exprimé par les deux équations A & B, la prémiére de l'ordre n & la seconde de l'ordre m.

$$\text{A} \ldots\ x^n - [1]x^{n-1} + [1^2]x^{n-2} - [1^3]x^{n-3} + \&c. \ldots [1^n] = 0$$

$$\text{B} \ldots\ (0)x^0 + (1)x^1 + (2)x^2 + (3)x^3 + \&c. \ldots + (m)x^m = 0$$

Les $1, 1^2, 1^3$, &c. dans des parenthéses quarrées, marquent, non les puissances de l'unité, mais les coëfficients de x, ou les fonctions rationnelles de y qui multiplient les puissances de x dans l'éq: A. Et les chiffres 0, 1, 2, 3, &c. dans les parenthéses rondes, indiquent aussi les fonctions rationelles d'y qui multiplient les puissances de x dans l'équation B. L'usage de cette Notation paroitra dans la suite, & les parenthéses ne laissent aucune équivoque. On propose de faire évanouir x, au moyen de ces deux équations, & de trouver celle qui exprime le raport des fonctions $[1]$, $[1^2]$, $[1^3]$, &c. (0), (1), (2), &c. c'est-à-dire de trouver l'équation en y & constantes, qui reste quand on a fait évanouir x. Nous nommerons cette équation C.

§. 2. Que a, b, c, d, &c. représentent les racines de l'éq: A, ou les fonctions, rationelles ou irrationelles, de y qui sont les valeurs de x dans cette équation $x^n - [1]x^{n-1}$ &c... $[1^n] = 0$. Comme elle est du dégré n, le nombre de ses racines est n. Et les substituant successivement dans l'éq: B, on aura n équations α, β, γ, δ, &c, où x ne paroit plus.

$$\alpha \ldots\ (0)a^0 + (1)a^1 + (2)a^2 \ldots\ldots + (m)a^m = 0$$
$$\beta \ldots\ (0)b^0 + (1)b^1 + (2)b^2 \ldots\ldots + (m)b^m = 0$$
$$\gamma \ldots\ (0)c^0 + (1)c^1 + (2)c^2 \ldots\ldots + (m)c^m = 0$$
$$\delta \ldots\ (0)d^0 + (1)d^1 + (2)d^2 \ldots\ldots + (m)d^m = 0$$

&c. &c.

Ces

Elimination of variables from polynomials, from Cramer (1750).

The roots of these equations are precisely the values of y that allow x to be eliminated from both (A) and (B), that is, they are the roots of an elimination equation (C). Thus, (C) must be simply the product of (α), (β), (γ), ... Cramer then used a lengthy combinatorial argument to prove what Euler had simply assumed, that the coefficients of (C) are always rational functions of the coefficients of (A) and (B). His reasoning, like Euler's more intuitive perception in 1748, depended on (i) the symmetric appearance of $a, b, c, \dots$ in each coefficient and (ii) the fact that

$$\begin{aligned} a + b + c + \cdots &= [1], \\ ab + ac + ad + \cdots + bc + bd + \cdots + cd + \cdots &= [1^2], \\ abc + abd + \cdots + bcd + \cdots &= [1^3], \\ &\ \ \cdots\cdot \end{aligned}$$

As a by-product Cramer's demonstration provided an algorithm for finding each coefficient, and he was able to show that his results agreed exactly with those given by Newton in the *Arithmetica universalis*.[13]

Cramer claimed that there were many useful consequences of this work but that he would give only the one he had aimed at, namely, that the degree of (C) can be no greater than mn and therefore it can have no more than mn roots.[14] From the point of view of later writers, however, his most important result was one that was from then on taken for granted: that the coefficients of an elimination equation can be expressed as rational functions of the coefficients of the original equations.

Euler's further thoughts on elimination, 1752

Shortly after the publication of Cramer's *Introduction à l'analyse des lignes courbes* in 1750, Euler returned once more to the problem of elimination. On 10 February 1752 he presented to the Berlin Academy a paper entitled 'Nouvelle méthode d'éliminer les quantités inconnues des equations' ('A new method of eliminating unknown quantities from equations').[15] On this occasion there was a particularly long delay between presentation and publication, so that the paper finally appeared in the *Mémoires* for 1764, printed in 1766.

In 1747 Euler had not mentioned Newton's elimination rules but now he did, prompted, perhaps, by Cramer's reference to them. In fact his first item was a demonstration of how he thought Newton had found his results. He began with the two equations

$$\begin{aligned} A + Bz &= 0, \\ a + bz &= 0, \end{aligned}$$

[13]Cramer 1750, 661–672.

[14]Cramer 1750, 672–676.

[15]The presentation is recorded in the Academy *Registre* for 1746–66, BBAW MS I–IV–31, f. 89v. A manuscript copy of the paper, now known as E310, is held by the Academy as BBAW MS I–M 116, C.7, 290–294.

where it is easy to see that there is a common root only if

$$Ab - Ba = 0.$$

Next Euler looked at a pair of quadratic equations. Thus, suppose we have

$$A + Bz + Czz = 0,$$
$$a + bz + czz = 0.$$

Multiplying the first of these by c and the second by C and subtracting one result from the other, we obtain

$$(Ac - Ca) + (Bc - Cb)z = 0.$$

Alternatively, multiplying the first by a and the second by A, we obtain

$$(Ba - Ab) + (Ca - Ac)z = 0.$$

Applying the rule for linear equations to these last two we have

$$(Ac - Ca)(Ca - Ac) - (Bc - Cb)(Ba - Ab) = 0.$$

Clearly one may continue in the same way for two cubics, two quartics, and so on. This was exactly the method Euler had proposed in the first part of his 'Demonstration' in 1748 (and in the second volume of his *Introductio* in the same year), and it did indeed confirm Newton's Rules I to IV.

Euler observed once again, however, that the method can produce redundant solutions; he had therefore been forced to consider the idea of elimination more carefully, in order to discern more precisely what it meant and what operations were needed in order to achieve it.[16] Thus Euler turned to a new approach, his 'nouvelle méthode' of the title.

To illustrate the new method Euler first took the equations (§12)

$$zz + Pz + Q = 0,$$
$$z^3 + pzz + qz + r = 0,$$

where, as before, P, Q, p, q, r, are functions of a second unknown, and considered the conditions under which these two equations can have a common root $z = w$. That is, he supposed that

$$zz + Pz + Q = (z - w)(Z + \mathfrak{A})$$

[16] *Or d'abord, l'idée de l'élimination ne paroissant pas asses précise, je commencerai par mieux déveloper cette idée, & par déterminer plus exactement, à quoi se réduit la question. Car, dès que nous nous seras formé une idée juste du sujet auquel aboutit l'élimination, nous verrons d'abord, quelles quéstions on sera obligé d'entreprendre par arriver à ce but. Thus he turned to a new approach, his 'nouvelle méthode'.* [Now first of all, since the idea of elimination does not appear to be sufficiently precise, I will begin by developing that idea better, and by determining more exactly what the question reduces to. For as soon as we have formed an exact idea on the subject of what elimination is, we will see straightaway what tasks we must undertake in order to arrive at it.] Euler (1764) [1766], §11.

and

$$z^3 + pzz + qz + r = (z - w)(zz + \mathfrak{a}z + \mathfrak{b})$$

for some $\mathfrak{A}$, $\mathfrak{a}$, $\mathfrak{b}$. This in turn led him to the equation

$$(zz + Pz + Q)(zz + \mathfrak{a}z + \mathfrak{b}) = (z + \mathfrak{A})(z^3 + pzz + qz + r)$$

and equating coefficients gave him

$$\begin{aligned} P + \mathfrak{a} &= p + \mathfrak{A}, \\ Q + P\mathfrak{a} + \mathfrak{b} &= q + p\mathfrak{A}, \\ P\mathfrak{b} + Q\mathfrak{a} &= q\mathfrak{A} + r, \\ Q\mathfrak{b} &= r\mathfrak{A}. \end{aligned}$$

These are linear in $\mathfrak{A}$, $\mathfrak{a}$, $\mathfrak{b}$, which can be eliminated to give an equation in P, Q, p, q, r, namely

$$Q(P-p)(Pq-Qp)+2Qr(P-p)+Pr(Q-q)-PPr(P-p)+Q(Q-q)^2+rr = 0.$$

In general given two equations (§16)

$$\begin{aligned} z^m + Pz^{m-1} + Qz^{m-2} + Rz^{m-3} + \cdots &= 0, \\ z^n + pz^{n-1} + qz^{n-2} + rz^{n-3} + \cdots &= 0, \end{aligned}$$

Euler could apply the same method, arriving at $m + n - 1$ linear equations from which $m + n - 2$ letters must be eliminated to give an equation in P, Q, R, ... and $p, q, r, \ldots$.

Euler admitted that his method had no particular advantage over some others as a method of elimination but argued that if was useful because it was easily applicable to certain problems in connection with curves. It could be used, for example, to discover where two equations shared repeated roots, which was useful if one wished to examine curves that intersected more than once for the same value of z. The method therefore had some practical value, but added little or nothing to the theory that he and Cramer had developed earlier.

Bezout's extension to more than two variables, 1764

Bezout's extension of elimination theory to more than two equations in more than two unknowns is not directly relevant to the problem of solving equations, but a brief account of his work is given here to show where the theory stood by the end of the 1760s. When Bezout wrote his paper, 'Rechérches sur le degré des équations résultantes de l'évanouissement des inconnues' ('Researches on the degree of equations resulting from the vanishing of unknowns') in the early 1760s, Euler's 'Nouvelle méthode' had not yet been published and Bezout knew only of Newton's findings from 1707, Euler's of 1748, and Cramer's of 1750, all of which he cited in his introduction. Newton, he said, had found some useful results but his method gave rise to superfluous roots (*racines inutiles*), and was laborious beyond the first few simple cases. Euler and Cramer had made improvements but only for two equations in two unknowns. Bezout admired their methods and said he would not have sought out others if theirs had been applicable to a greater number of equations. He pointed out that even for three equations of degree 3, by no means the most troublesome case one can imagine, eliminating unknowns from two equations at a time will lead to an elimination equation of degree 81, even though one can see that the degree need be no more than 49 (though he did not explain how one can see that).

The chief difficulty, according to Bezout, lay precisely in detecting the superfluous factors. Even for just two equations, he said, the task might defeat any but the most intrepid calculator but should in principle be possible. The same was not true, however, where there were more than two equations, where one might search in vain, the only hope being to return to comparing two equations at a time. What was the thread, Bezout asked, that could guide one through such a labyrinth? He believed that so far there was neither any certain way of finding an elimination equation of the correct degree or even of determining what that degree should be. These were the problems he now proposed to tackle.

Bezout's paper is in two parts. In the first he derived some particular results for the degree of the elimination equation for two, three, four, or five equations in two, three, four, or five unknowns.[17] His results were not easy to apply, however, and the only specific example he gave was for the two equations

$$a^3x^5y - 2a^4y^2x^3 + y^8x - a^9 = 0,$$

$$a^3x^3 - 3a^3xy^2 + y^5x - y^6 = 0,$$

for which, according to Bezout, the degree of the elimination equation should be 42 (he did not say how he arrived at that). The second part of his paper contains his efforts to streamline the elimination procedure.[18] For the equations

$$Ax^m + Bx^{m-1} + Cx^{m-2} + \cdots + T = 0, \tag{15}$$

$$A'x^m + B'x^{m-1} + C'^{m-2} + \cdots + T' = 0, \tag{16}$$

for instance, he suggested that one should (i) multiply (15) by A' and (16) by A and subtract one result from the other to obtain an equation of degree $m-1$; (ii) multiply (15) by $A'x+B'$ and (16) by $Ax+B$ and subtract the results to obtain another equation of degree $m-1$; (iii) multiply (15) by $A'x^2+B'x+C'$ and (16) by Ax^2+Bx+C, and so on. From the m equations of degree $m-1$ found in this way it should be possible to eliminate powers of x and discover the necessary relationships between $A, B, C, \ldots, T$ and $A', B', C', \ldots, T'$. Such work was based partly on suggestions

[17] Bezout (1764) [1767], 301–317.
[18] Bezout (1764) [1767], 317–337.

made by Euler at the end of his chapter on elimination in the *Introduction ad analysin infinitorum*.[19]

Bezout complained more than once that for all but the simplest cases his methods became long and wearisome, and he ended his paper hoping that now he had given some indications others would continue to develop them. In fact he himself was to devote a great deal more attention to this subject in the coming years, resulting in the publication in 1779 of the work for which he is now best known, his *Théorie générale des équations algébriques*. Bezout's 1764 paper, eventually published in 1767, added significantly to the small but increasing number of writings on elimination theory and helped to establish it as a subject worthy of study. One of the people who noted this accumulation of papers in the mid 1760s was Lagrange.

Lagrange's ideas on elimination, 1769 and 1771

Lagrange, like Bezout, was a longstanding admirer and careful reader of Euler. As early as 1755 when he was only nineteen, Lagrange, then living in Turin, had sent his early mathematical writings to Euler in Berlin. Euler was so impressed that he proposed Lagrange as a member of the Berlin Academy, to which Lagrange was elected in 1756. Euler also tried, as did Maupertuis and later d'Alembert, to persuade Lagrange to take up a post at the Academy, but Lagrange modestly and persistently refused. In the end, he moved to Berlin at the personal invitation of Frederick II only after Euler left for St Petersburg in 1766, so that he and Euler never met in person. Mathematically, however, he was Euler's closest and most gifted follower.

It seems that the appearance of some of Euler's thoughts on elimination theory in 1766, closely followed by Bezout's paper in 1767, encouraged Lagrange too to give the matter some attention. He presented his first paper on the subject, entitled 'L'élimination des inconnues dans les équations' ('The elimination of unknowns in equations'), to the Berlin Academy in October 1767, but the usual publishing delays meant that it appeared in the volume of *Mémoires* for 1769, which was not printed until 1771. By that time, Lagrange had gone on to do much more detailed work on equations, and only a brief summary of his results of 1767 is needed here.

The problem of eliminating an unknown from two equations, as was by then well known, was that it could lead to an equation of degree higher than one actually needs. Lagrange, always keenly aware of his predecessors, referred to the early methods of Euler (1748) [1750] and Cramer (1750) as well as to the more recent proposals by Euler (1764) [1766] and Bezout (1764) [1767] for circumventing this difficulty. His aim now was to add a further method that offered general and easy rules. In outline his method was as follows.

Suppose we have two equations, of degree m and n respectively, from which x is to be eliminated:

$$1 + Ax + Bx^2 + Cx^3 + \cdots = 0, \tag{17}$$

[19]Euler, *Introductio ad analysin infinitorum*, 1748, II, §483–§485.

$$1+\frac{a}{x}+\frac{b}{x^2}+\frac{c}{x^3}+\cdots=0. \tag{18}$$

Lagrange said nothing about the nature of $A, B, C, \ldots$ or $a, b, c, \ldots$. He assumed, however, that the roots of (18) were $\frac{1}{\alpha}, \frac{1}{\beta}, \frac{1}{\gamma}, \ldots$, and eventually arrived, as had Euler and Cramer, at the elimination equation

$$\begin{aligned}\Pi &= (1+a\alpha+b\alpha^2+c\alpha^3+\cdots)\\ &\quad\times(1+a\beta+b\beta^2+c\beta^3+\cdots)\\ &\quad\times(1+a\gamma+b\gamma^2+c\gamma^3+\cdots)\\ &\quad\times\cdots\\ &= 0.\end{aligned}$$

The problem here, as Euler had already observed, is that we do not know $\alpha, \beta, \gamma, \ldots$ individually. Lagrange's solution to this problem was to write $\log \Pi$ as the sum of the logarithms of the factors on the right, for each of which he could write down a power series expansion. This led him eventually to the equation

$$\log \Pi = -\phi = -pP - 2qQ - 3rR - \cdots$$

where $P, Q, R, \ldots$ are sums of powers of reciprocals of the roots of (17) and $p, q, r, \ldots$ are sums of powers of roots of (18). In this way he arrived at the beautifully simple equation

$$\Pi = \mathrm{e}^{-\phi}.$$

In practice, despite various shortcuts suggested by Lagrange, the calculation of ϕ and $\mathrm{e}^{-\phi}$ is no easier than tackling the elimination with bare hands and Lagrange gave worked examples only for $m = n = 2$ and $m = n = 3$, cases that had already been well explored.

Two years later he offered a simpler outline of the problem in his lengthy 'Réflexions sur la résolution algébriques des équations', presented to the Academy in 1771 and published in 1772. This paper will be discussed at length in Chapter 10 and so only the two paragraphs on elimination are noted here. In §12 and §13 of the paper Lagrange supposed that two equations in x are represented by

$$P = 0,$$
$$Q = 0.$$

As in 'L'élimination des inconnues' in 1769 he said nothing about the nature of the coefficients. Nevertheless, he claimed, as had Euler in 1748 and Cramer in 1750, that if the roots of $Q = 0$ are $x', x'', x'', \ldots$, then the elimination equation is formed by constructing the product

$$P(x')P(x'')P(x''')\ldots \tag{19}$$

and then setting this equal to 0.

Lagrange went on to claim that the product in (19) can be found without actually solving for $x', x'', x''', \ldots$. It was easy to convince oneself of this, he argued, by noting that the product is unchanged by permutations of x', x'', x''', ..., that is to say, by permutations of $P(x')$, $P(x'')$, $P(x''')$, ... Further, since $P(x')$, $P(x'')$, $P(x''')$, ... all depend on $x', x'', x''', \ldots$ in a similar way, the functions of $x', x'', x''', \ldots$ that constitute the product (19) will be what we would now describe as 'symmetric'. Therefore, Lagrange claimed, they will be expressible in terms of the coefficients of the equation $Q = 0$ alone, without solving for x', x'', x''', ... individually.[20] He offered no proof of this statement. To him it appears to have been self-evident, based on his reading of Euler and Cramer and on his findings earlier in his own paper. For the technicalities of calculating the product he referred his readers to Cramer's *Introduction à l'analyse des lignes courbes* and to his own work in 'L'élimination des inconnues'.

Summary

The beginnings of elimination theory, as of so many of the stories in Part II of this book, were already hinted at in Newton's *Arithmetica universalis*, with its rules for elimination in a few special cases. As also so often happened, it was Euler who took the next important step. Euler's early work on elimination appears to have been independent of Newton's, motivated instead by his investigations into intersections of curves, a matter that was being explored around the same time, with similar results, by Cramer. Only in the early 1750s did Euler explicitly comment on, explain, and extend Newton's rules, though this work did not appear in print until 1766.

Meanwhile, Bezout too had taken up Newton's rules together with the early work of Euler and Cramer; his developments of the theory were published in 1767, just a year after Euler's later paper. It would seem that it was this near simultaneous publication of papers that drew in Lagrange, who within a short time was to bring together all the disparate strands of contemporary work on equations. Before turning more fully to Lagrange, however, we need first to explore a difference of opinion between Euler and Bezout.

[20] *on trouvera toujours que les différentes fonctions de x', x'', x''' &c. qui entreront dans le produit total seront exprimable par les seuls coëfficients de l'équation $Q = 0$, dont x', x'', x''' &c. sont les racines;* Lagrange (1770) [1772], §13.

Chapter 8
The degree of a resolvent

In his 'Mémoire sur plusieurs classes d'équations [...] qui admettent une solution algébrique', written in 1762, Bezout remarked in passing on a hypothesis of Euler's with which he disagreed. Bezout's work on Descartes' method for quartic equations had demonstrated that the resolvent equation or *la réduite*, as he himself always called it, must be of degree 6. On similar grounds, he predicted that a resolvent for a fifth-degree equation would be of degree 10; and that the degree of a resolvent would in general be higher than the degree of the original equation. Euler in 1733 had come to a different conclusion. The well known Cartesian resolvent of degree 6 for a quartic contains only even powers of the unknown and so can in fact be solved as a cubic. Similarly, the resolvent for a cubic, though also of degree 6, contains only third and sixth powers, and so can be solved as a quadratic. Such results had led Euler to suggest that there would always be a resolvent of degree one less than the degree of the original (see page 110), a much more comforting suggestion than Bezout's.

To some extent this disagreement had already been anticipated in the seventeenth century: Hudde, Gregory, and Leibniz had all discovered that trying to solve an equation of degree 5 led them to an equation of degree 10 or even 20, whereas Tschirnhaus seems to have remained convinced that it should be possible to reduce the degree of any equation by one (see page 65). Euler and Bezout knew nothing of the private deliberations of Gregory, Leibniz, and Tschirnhaus, and if either had read Hudde's derivation of an equation of degree 10 they did not mention it. The question therefore seems to have arisen afresh for them as their research into the structure of equations began to deepen.

Having noted the discrepancy between his hypothesis and Euler's, Bezout continued to work on the problem, and presented a *précis* of his further findings to the Paris Academy in January 1763. The time needed to complete the work, however, together with the usual publication delays, meant that his finished paper, 'Mémoire sur la résolution générale des équations de tous les degrés' ('Memoir on the general solution of equations of any degree') was not included in the Academy Memoirs until 1765 and did not appear in print until 1768. The memoir proved to be enormously influential and so is described in this chapter in some detail.

Bezout began with a *résumé* of current knowledge.[1] To give a general solution of an equation, he stated, is to give algebraic expressions for each of its roots in terms of the coefficients. Further, such expressions can contain radicals of every degree up to and including the degree of the equation. Bezout's justification for this claim was that the solution of an equation of degree n can include at least one n^{th} root, that is, a radical of degree n. If the constant term of the equation is zero, however, the degree

[1] Bezout (1765) [1768], 534–536.

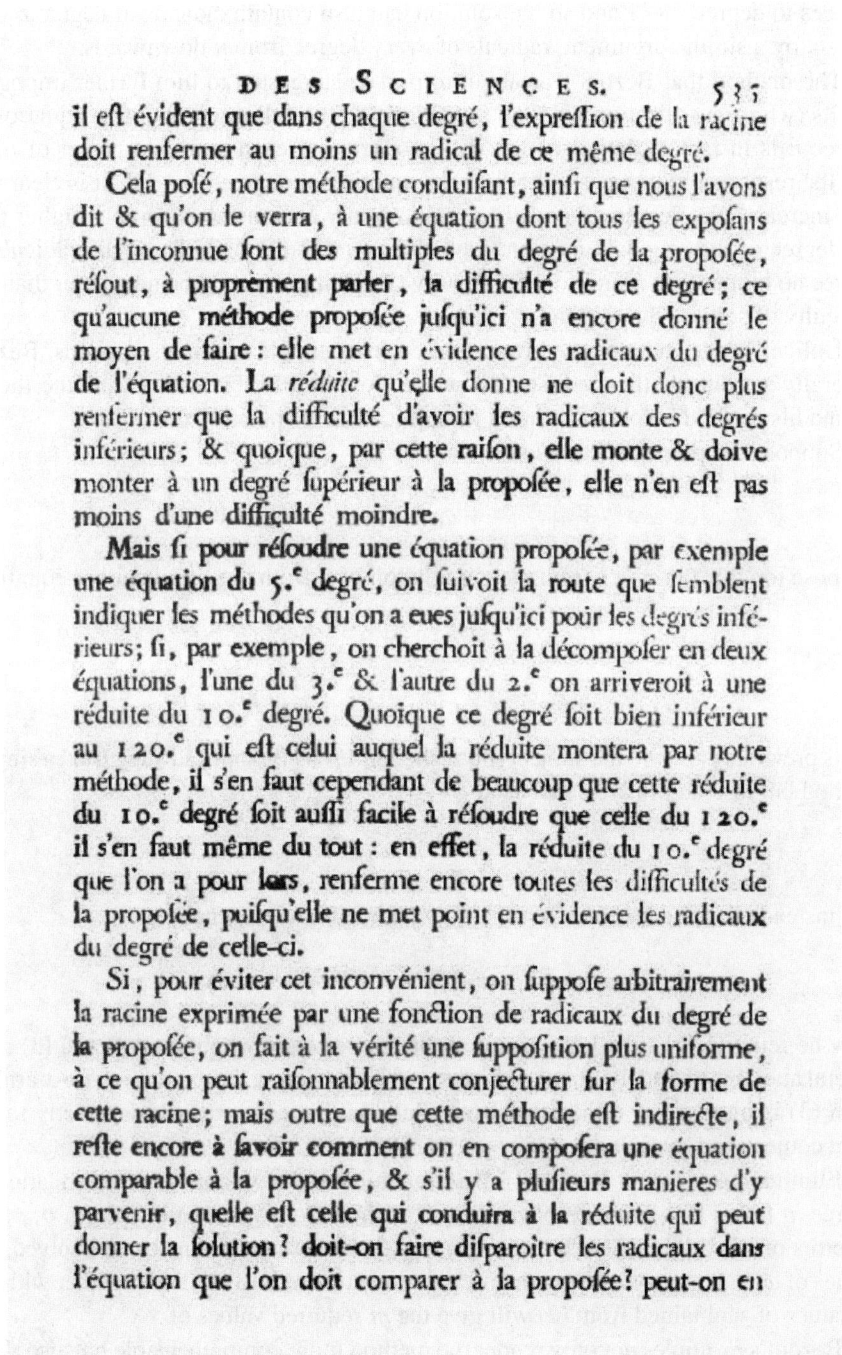

DES SCIENCES. 535

il eſt évident que dans chaque degré, l'expreſſion de la racine doit renfermer au moins un radical de ce même degré.

Cela poſé, notre méthode conduiſant, ainſi que nous l'avons dit & qu'on le verra, à une équation dont tous les expoſans de l'inconnue ſont des multiples du degré de la propoſée, réſout, à proprèment parler, la difficulté de ce degré; ce qu'aucune méthode propoſée juſqu'ici n'a encore donné le moyen de faire: elle met en évidence les radicaux du degré de l'équation. La *réduite* qu'elle donne ne doit donc plus renfermer que la difficulté d'avoir les radicaux des degrés inférieurs; & quoique, par cette raiſon, elle monte & doive monter à un degré ſupérieur à la propoſée, elle n'en eſt pas moins d'une difficulté moindre.

Mais ſi pour réſoudre une équation propoſée, par exemple une équation du 5.e degré, on ſuivoit la route que ſemblent indiquer les méthodes qu'on a eues juſqu'ici poür les degrés inférieurs; ſi, par exemple, on cherchoit à la décompoſer en deux équations, l'une du 3.e & l'autre du 2.e on arriveroit à une réduite du 10.e degré. Quoique ce degré ſoit bien inférieur au 120.e qui eſt celui auquel la réduite montera par notre méthode, il s'en faut cependant de beaucoup que cette réduite du 10.e degré ſoit auſſi facile à réſoudre que celle du 120.e il s'en faut même du tout: en effet, la réduite du 10.e degré que l'on a pour lors, renferme encore toutes les difficultés de la propoſée, puiſqu'elle ne met point en évidence les radicaux du degré de celle-ci.

Si, pour éviter cet inconvénient, on ſuppoſe arbitrairement la racine exprimée par une fonction de radicaux du degré de la propoſée, on fait à la vérité une ſuppoſition plus uniforme, à ce qu'on peut raiſonnablement conjecturer ſur la forme de cette racine; mais outre que cette méthode eſt indirecte, il reſte encore à ſavoir comment on en compoſera une équation comparable à la propoſée, & s'il y a pluſieurs manières d'y parvenir, quelle eſt celle qui conduira à la réduite qui peut donner la ſolution? doit-on faire diſparoître les radicaux dans l'équation que l'on doit comparer à la propoſée? peut-on en

Discussion of equations of degree 5, from Bezout (1765).

reduces to degree $n-1$ and so the solution can also contain radicals of degree $n-1$; and so, by a similar argument, radicals of every degree from n downwards.

The method that Bezout was about to pursue suggested to him further important results (whose justification we shall see shortly): (i) a full resolvent of an equation of degree n is in fact of degree $n!$ but (ii) the degree of each term is a power of n, so that the resolvent reduces essentially to an equation of degree $(n-1)!$. It is clear that as n increases the degree of a resolvent will rapidly become considerably higher than the degree of the original equation; nevertheless, its solution will contain radicals of degree no higher than n and so the difficulty of solving it should be no greater than the difficulty of solving the original.

Unlike Euler, who almost always worked from simple examples upwards, Bezout generally set out his theory first. We will follow him in this and outline the theory behind his method before examining its application to specific examples.[2]

Suppose the proposed equation is

$$x^m + px^{m-2} + qx^{m-3} + \cdots + T = 0. \tag{1}$$

Suppose too that (1) arises from the elimination of y from the simultaneous equations

$$y^m - 1 = 0, \tag{2}$$

$$ay^{m-1} + by^{m-2} + cy^{m-3} + \cdots + h + x = 0. \tag{3}$$

In his previous paper on the subject (published in 1764) Bezout had suggested a similar method but instead of (2) he had used

$$y^m + h = 0; \tag{2a}$$

and instead of (3) he had used $y = \frac{x+a}{x+b}$, which led him eventually to

$$x = b(y^{n-1} + y^{n-2} + y^{n-3} + \cdots + y). \tag{3a}$$

Now he remarked that he had arrived at the revised transformations (2) and (3) after several attempts to find the simplest forms possible. Clearly the expression for x arising from (3) is more general than that from (3a). In fact (3) corresponds exactly to the form conjectured by Euler in 1753 (see page 113).

Eliminating y from (2) and (3), Bezout claimed, we will arrive at an equation of degree m in x. This can then be compared with (1) to give equations for $a, b, c, \ldots$, in terms of the original coefficients $p, q, r, \ldots$. If these equations can be solved, the values of $a, b, c, \ldots$ found from them can be substituted into (3), and together with the m values of y obtained from (2) will give the m required values of x.

Bezout's examples not only render the method more comprehensible but also show up some important results. He applied his method first to cubics and then to quartics,

[2]Bezout (1765) [1768], 536–537.

and here we will examine both.[3] In order to relate the examples to the theory and to each other, I have numbered equations equivalent to (1) as (1′), (1″), and so on.

Suppose we wish to solve the cubic equation

$$x^3 + px + q = 0. \tag{1′}$$

Bezout's method instructs us to eliminate y from the two equations

$$y^3 - 1 = 0, \tag{2′}$$

$$ay^2 + by + x = 0. \tag{3′}$$

Multiplying (3′) by 1, y, y^2, respectively, (and using the fact that $y^3 = 1$) gives

$$\begin{aligned} ay^2 + by + x &= 0, \\ by^2 + xy + a &= 0, \\ xy^2 + ay + b &= 0. \end{aligned}$$

From these equations (linear in y and y^2) we can eliminate y and y^2 in the usual way to arrive at

$$x^3 - 3abx + (a^3 + b^3) = 0. \tag{4′}$$

Comparing (4′) with (1′) gives

$$\begin{aligned} -3ab &= p, \\ a^3 + b^3 &= q, \end{aligned}$$

and hence the usual equation for a (or b), namely,

$$a^6 - qa^3 - \tfrac{1}{27}p^3 = 0. \tag{5′}$$

Thus (5′) is a resolvent for (1′). It is an equation of degree 6 (or 3!) but the degree of each term is a multiple of 3, so that its 'difficulty' (*difficulté*), as Bezout put it, is only of degree 2. Each value of a that satisfies (5′) gives rise to a single corresponding value of b because of the relationship $-3ab = p$. Any such pair of values of a and b can be substituted into (3′) (Bezout did not comment on which of the six possible pairs one should choose). The three values of y given by (2′) and substituted into (3′) then yield the three required solutions of (1′).

Bezout's next example was the solution of the quartic

$$x^4 + px^2 + qx + r = 0. \tag{1″}$$

The two equations from which y must now be eliminated are

$$y^4 - 1 = 0, \tag{2″}$$

[3] Bezout (1765) [1768], 537–540.

$$ay^3 + by^2 + cy + x = 0. \tag{3''}$$

Multiplying (3″) by 1, y, y^2, y^3 respectively, gives rise to the four equations

$$\begin{aligned} ay^3 + by^2 + cy + x &= 0, \\ by^3 + cy^2 + xy + a &= 0, \\ cy^3 + xy^2 + ay + b &= 0, \\ xy^3 + ay^2 + by + c &= 0, \end{aligned}$$

from which y, y^2, y^3 can be eliminated to give

$$x^4 - (4ac + 2b^2)x^2 + (4a^2b + 4bc^2)x - (a^4 + c^4 - b^4 - 2a^2c^2 + 4ab^2c) = 0. \tag{4''}$$

Comparison of (4″) with (1″) gives equations that Bezout labelled (A), (B), and (C):

$$4ac + 2b^2 = -p, \tag{A}$$

$$4a^2b + 4bc^2 = q, \tag{B}$$

$$a^4 + c^4 - b^4 - 2a^2c^2 + 4ab^2c = -r. \tag{C}$$

Bezout claimed (and later demonstrated) that if one solves (A), (B), and (C) for either a or c one arrives at an equation of degree 24 (or 4!). However, the degree of each term is a multiple of 4, and so the equation reduces essentially to degree 6. If instead, however, one solves (A), (B), and (C) for b, the resulting equation is immediately of degree 6 and contains only even powers, and is therefore solvable as an equation of degree 3. Once a value of b can be found, so are corresponding values of a and c, which can be substituted into (3″). The four possible values of y from (2″), namely, $1, -1, \sqrt{-1}, -\sqrt{-1}$, then give the four required solutions to (1″).

One can see why it was that around this time Bezout became particularly interested in the degree of the elimination equation when there are more than two equation in more than two unknowns (see page 141–143).

After working through these examples Bezout embarked upon some reflections.[4] One might think, he ventured, like some analysts (clearly he was thinking of Euler), that a quintic could be solved with the aid of an equation of degree 4, that is, that the resolvent will be of degree 20 but with the degree of each term a multiple of 5. The examples just demonstrated, however, cast doubt on this. Bezout thought there was a very great probability (*une très-grande probabilité*) that the resolvent would be of degree 120 (or 5!). The reduction in degree from 24 to 6 in the case of quartics, he thought, was an 'accidental simplification' (*simplification accidentelle*) that comes about in the following way.[5]

Consider again the problem of solving equations (A), (B), and (C). We can see that they are symmetric with respect to a and c or, as Bezout put it, b is 'disposed in the

[4] Bezout (1765) [1768], 540–543.

[5] Bezout (1765) [1768], 541–542.

same way' (*disposée de la même manière*) with regard to these letters. Thus, he argued, an equation for b is bound to be simpler than an equation for either a or c. The special position of b arises, Bezout explained, from the fact that it is the middle coefficient of the three in (3″) and is therefore in the same 'disposition' to each of a and c. A similar symmetry will arise in any equation of even degree, but not for equations of odd degree. For a quintic, for example, the equation corresponding to (3) is

$$ay^4 + by^3 + cy^2 + dy + x = 0,$$

which has four coefficients a, b, c, d, none of which can be said to take preference over the others.

Besides, asked Bezout, how can we think that the coefficient a, which turns out to take 6 values for cubics and 24 for quartics, can possibly take only 20 values for quintics? What law could determine an outcome so bizarre? (*quelle seroit la loit qui règleroit une marche aussi bizarre*)? Bezout insisted that the fault did not lie in his method, which gave better results than any other he knew. Further, because it was applicable to equations of any degree, it showed 'by analogy' what the degree of the resolvent of any equation must be, as well as some of the special cases where the degree may be reduced.

Bezout's conclusions, already outlined in his opening paragraphs, were based in particular on his findings for quartic equations, where a full resolvent is of degree 4! but where the degree of each term is a multiple of 4 so that it reduces essentially to degree 3!, that is, it can be solved using only square and cube roots. On these grounds, Bezout claimed, there is good reason to think that (i) for an equation of degree n the degree of each term of a full resolvent will be a multiple of n and therefore the latter will reduce to an equation of lower degree but (ii) although the solution of the resolvent will depend only on lower degree equations it will involve a combination of all the 'difficulties' of such equations.

From here Bezout passed to his final example, a general quintic of the form

$$x^3 + 5px^3 + 5qx^2 + 5rx + s = 0, \qquad (1''')$$

which he wished to compare with the elimination equation of

$$y^5 - 1 = 0 \qquad (2''')$$

and

$$ay^4 + by^3 + cy^2 + dy + x = 0. \qquad (3''')$$

Bezout went through exactly the same procedure as for cubics and quartics, but the resulting equation (4‴) of degree 5 in x is, of course, very much more unwieldy than (4′) or (4″). It turns out that r, for instance, is the sum of seven terms each of degree 4 in a, b, c, d; while s is the sum of twelve terms each of degree 5. It is clearly extremely difficult, if not impossible, to solve such equations except in special cases; for example, where any two of a, b, c, d are zero. Such special cases were precisely those that Bezout had already discovered and written about in his previous paper (1762) [1764].

Where the degree of the original equation is composite, say kl, Bezout offered a refinement which, he claimed, led to easier calculations. Instead of equations (2) and (3) above he now took

$$y^k - 1 = 0$$

and

$$\begin{aligned} &y^{k-1}(ax^{l-1} + bx^{l-2} + \cdots + h) \\ &\quad + y^{k-2}(a'x^{l-1} + b'x^{l-2} + \cdots + h') \\ &\qquad + y^{k-3}(a''x^{l-1} + b''x^{l-2} + \cdots + h'') \\ &\qquad\quad + \cdots + x^l + Ax^{l-1} + Bx^{l-2} + \cdots + P = 0. \end{aligned}$$

After this the method proceeds as before, and Bezout worked it in detail for $kl = 4$ and $kl = 6$, but it added nothing to the insights he had described earlier.

Summary

In the seventeenth century Hudde, Gregory, and Leibniz had all discovered that trying to solve an equation of degree 5 led them to equations of higher degree, 10 or even 20. In 1733, Euler, unaware of their work and thus of the lessons of history, thought that it should always be possible to reduce the degree of an equation by one. Unfortunately, his habitual method of generalizing from easy cases had for once led him badly astray: the reductions that are possible for cubic and quartics become much more elusive for quintics, but Euler failed to investigate equations of degree 5 carefully enough to draw the correct conclusions and remained convinced that the difficulty lay chiefly in the calculations.

Bezout in 1762 was the first to see that arriving at equations of higher degree was not just a quirk of any particular method but a deep-rooted problem: that the degree of a resolvent would in general always be higher than the degree of the original equation. He was not only able to argue convincingly against Euler's hypothesis but also suggested that in general an equation of degree n would give rise to a resolvent of degree $n!$, or perhaps $(n-1)!$. In either case the difficulty of solving it was going to be far greater than that of solving the original.

Chapter 9
Numerical solution

The aim of most of the mathematics discussed in this book so far was to discover rules or procedures that would deliver the roots of polynomial equations from their coefficients. Even for equations of degree three or four the calculations could be formidable, while for equations of degree higher than that the problem was proving stubbornly intractable. A method of finding a numerical approximation to a root, was therefore not only desirable but essential, as Viète had recognized as long ago as 1591 (see pages 29–31). Further, the benefits of such methods went beyond the merely practical: efforts to understand and improve numerical techniques could lead to new insights into the structure and properties of equations, as we saw in the work of Harriot (pages 35–42).

In the early seventeenth century, Viète's method was known as the 'general way' (*via generalis*), of solving equations. It was taken up not only by Harriot soon after 1600 but also by another English admirer of Viète, William Oughtred, in an appendix to *The key of mathematics* in 1647 (the first English edition of his *Clavis mathematicae* of 1631).[1] It was revived in the nineteenth century by William George Horner and became known as the Horner method. In this chapter we examine two other methods of numerical solution, proposed by Newton and Lagrange, respectively. Newton's method was a by-product of other research: he was never particularly interested in numerical solution for its own sake. Lagrange, on the other hand, focussed very specifically on the problem, and in doing so drew upon a great deal of the work that has been described earlier in this book.

Newton's iteration method, 1660s

During the 1660s, in the course of his early research on infinite series, Newton discovered a method of numerical solution based on his insight that a decimal expansion is in essence a power series in decreasing powers of 10. Thus the method he devised to elicit solutions of literal equations as power series could also be used to calculate roots of numerical equations, digit by digit. Newton wrote out two detailed examples in 1669 in 'De analysi', his first written treatise on infinite series.[2] 'De analysi' remained unpublished until 1711, but Newton's examples became well known because he sent them to Leibniz in June 1676,[3] and they were subsequently published by Wallis in *A treatise of algebra* in 1685.[4]

---

[1]Oughtred 1647, 139–169.

[2]Newton 1967–81, II, 218–233.

[3]Newton to Oldenburg for Leibniz, 13 June 1676 (*Epistola prior*), in Newton 1959–77, II, 23–24 and 34–35. Turnbull's translation is misleading: by *Extractiones Radicum affectarum* Newton did not mean 'extractions of affected roots' but 'extractions of roots of affected equations'.

[4]Wallis 1685, 339–340.

Newton's first example was the equation $y^3 - 2y - 5 = 0$. By inspection it can be seen that the best integer approximation to the single real root is $y = 2$. Newton claimed that this differs from the true root by less than a tenth part, namely, $\frac{2}{10}$, though this is not obvious without further calculation. Now he put $y = 2 + p$, so that p must satisfy

$$(2 + p)^3 - 2(2 + p) - 5 = 0,$$

that is,

$$p^3 + 6p^2 + 10p - 1 = 0.$$

Since p is supposed small in relation to 2 (less than one tenth of it), p^2 and p^3 may be neglected, and so we have the estimate $10p - 1 = 0$ or $p = 0.1$. Next Newton refined this by putting $p = q + 0.1$ and repeating the procedure to give

$$q^3 + 6.3q^2 + 11.23q + 0.061 = 0$$

and thus an estimated value of q:

$$q = -\frac{0.061}{11.23} = -0.0054.$$

After one further step, in which he put $q = r - 0.0054$, Newton obtained $r = -0.00004853$ 'nearly' (*proxime*), and consequently $y = 2{+}0.1{-}0.0054{-}0.0004853 = 2.09455147$.

Newton set out his calculations in tabular form as shown here:

$2 + p = y$	y^3	$+8$	$+12p$	$+6p^2$	$+p^3$
	$-2y$	-4	$-2p$		
	-5	-5			
	Summa	-1	$+10p$	$+6p^2$	$+p^3$
$0.1 + q = p$	$+p^3$	$+0.001$	$+0.03q$	$+0.3q^2$	$+q^3$
	$+6p^2$	$+0.06$	$+1.2$	$+6.0$	
	$+10p$	$+1 + 10$			
	-1	-1			
	Summa	$+0.061$	$+11.23q$	$+6.3q^2$	$+q^3$
$-0.0054 + r = q$	$6.3q^2$	$+0.000183708$	$-0.06804r$	$+6.3r^2$	
	$+11.23q$	-0.060642	$+11.23$		
	0.061	$+0.061$			
	Summa	$+0.000541708$	$+11.1619r$	$+6.3r^2$	
-0.00004853					

PER ÆQUATIONES INFINITAS. 9

Et ſupponens $-0{,}0054 + r = q$, hunc ut prius ſubſtituo, & operationem ſic produco quo uſq; placuerit. Verum ſi ad bis tot figuras tantum quot in Quotiente jam reperiuntur una dempta, operam continuare cupiam, pro q ſubſtituo $-0{,}0054 + r$ in hanc $6{,}3q^2 + 11{,}23q + 0{,}061$, ſcilicet primo ejus termino (q^3) propter exilitatem ſuam

$y^3 - 2y - 5 = 0$		$+2{,}10000000$ $-0{,}00544853$ $+2{,}09455147 = y$
$2 + p = y$	$+y^3$	$+8 + 12p + 6p^2 + p^3$
	$+2y$	$-4 - 2p$
	-5	-5
	Summa	$-1 + 10p + 6p^2 + p^3$
$0{,}1 + q = p$	$+p^3$	$+0{,}001 + 0{,}03q + 0{,}3q^2 + q^3$
	$+6p^2$	$+0{,}06 \quad + 1{,}2 \quad + 6{,}0$
	$+10p$	$+1, \quad + 10,$
	-1	$-1,$
	Summa	$+0{,}061 + 11{,}23q + 6{,}3q^2 + q^3$
$-0{,}0054 + r = q$	$+6{,}3q^2$	$+0{,}000183708 - 0{,}06804r + 6{,}3r^2$
	$+11{,}23q$	$-0{,}060642 \quad + 11{,}23$
	$+0{,}061$	$+0{,}061$
	Summa	$+0{,}000541708 + 11{,}16196r + 6{,}3r^2$
$-0{,}00004854 + s = r$		

neglecto, & prodit $6{,}3r^2 + 11{,}16196r + 0{,}000541708 = 0$ fere, ſive (rejecto $6{,}3r^2$) $r = \frac{-0{,}000541708}{11{,}16196} = -0{,}00004853$ fere, quam ſcribo in negativa parte Quotientis. Denique negativam partem Quotientis ab Affirmativa ſubducens habeo $2{,}09455147$ Quotientem quæſitam.

Æquationes plurium dimenſionum nihilo ſecius reſolvuntur, & operam ſub fine, ut hic factum fuit levabis, ſi primos ejus terminos gradatim omiſeris.

Præterea notandum eſt quod in hoc exemplo, ſi dubitarem an $0{,}1 = p$ veritati ſatis accederet, pro $10p - 1 = 0$, finxiſſem $6p^2 + 10p - 1 = 0$, & ejus radicis primam figuram in Quotiente ſcripſiſſem; & ſecundam vel tertiam Quotientis figuram ſic explorare convenit, ubi in Æquatione iſta ultimo reſultante quadratum coefficientis penultimi termini, non ſit decies majus quam factus ex ultimo termino ducto in coefficientem termini antepenultimi.

C Imo

A method of solving equations numerically, from Newton (1711).

In 'De analysi' Newton followed this numerical example with the literal equation $y^3 + a^2y - 2a^3 + axy - x^3 = 0$, which he solved by exactly the same procedure to obtain

$$y = a - \frac{x}{4} + \frac{x^2}{64a} + \frac{131x^3}{512a^2} + \frac{509x^4}{16384a^3} + \cdots .$$

This easy transition from numerical to literal examples was typical of Newton's handling of power series.

Newton regarded his method as intuitive and easy to remember, and indeed it is.[5] The difficulty is not in understanding the procedure, but in knowing when it will work, and where to begin the iteration. For literal equations Newton offered some guidance on this last question, using what came to be known as his 'algebraic parallelogram'.[6] Terms that could possibly appear in an equation are arranged in a rectangular grid, and those that actually appear are marked with $*$. Thus for the equation $y^3 + a^2y - 2a^3 + axy - x^3 = 0$ Newton's grid was:

$*x^3$	x^3y	x^3y^2	x^3y^3
x^2	x^2y	x^2y^2	x^2y^3
x	$*xy$	xy^2	xy^3
$*1$	$*y$	y^2	$*y^3$

He then drew a straight line across the grid below the lowest entry in the left hand column and below any subsequent starred entries. Entries not adjacent to the line are then temporarily neglected. For the equation above, $y^3 + a^2y - 2a^3 + axy - x^3 = 0$, containing terms in y^3, y, 1, xy, x^3, the line runs horizontally below $*1$, $*y$, and $*y^3$. The equation may therefore be temporarily reduced to $y^3 + a^2y - 2a^3 = 0$, with solution $y = a$. This solution, Newton claimed, may be taken as the starting point of the iteration. The parallelogram does not help, however, to find starting values for numerical equations, and for these Newton offered no suggestions except to find the nearest integer by trial and error. His neglect of this question was probably due to the fact that solving numerical equations was not his main concern: the method came out of his deeper research into infinite series where he had other methods of finding a starting value.

During the late 1660s Newton laid the foundations of his *Enumeratio curvarum trium dimensionum*, his classification of cubic curves. In the published version (appended to his *Opticks* in 1704) he said little about the methods he had used, but everyone who later commented on it assumed that it rested on Newton's ability to derive infinite

---

[5] *Demonstratio ejus ex ipso modo operandi patet, unde cum opus sit in memoriam facile revocatur.* [The justification of this [method] is clear from the way of working, from which it is easily called to mind when needed.] Newton 1967–81, II, 222.

[6] Newton 1959–77, 126–127 and 145–146; see also Wallis 1685, 339–340.

series solutions to algebraic equations.[7] Eighteenth-century writers continued to use the method for algebraic purposes.[8] It was his numerical method, however, that was more rapidly taken up. Very soon after it first appeared in print in Wallis's *Treatise of algebra*, a simplified general version of it was described by Joseph Raphson in his *Analysis aequationum universalis* (*General analysis of equations*) (1690).[9]

Details of Raphson's life are surprisingly obscure. His date of birth is often given as 1648, but this seems to be quite wrong since Edmund Halley referred to him in 1694 as a young man.[10] Raphson's name and his knowledge of the cabbala, as demonstrated in his *Demonstratio de deo* (1710), have led David Thomas to suggest that he was of Jewish origin and from an Irish immigrant family.[11] He was admitted to the Royal Society late in 1689, possibly on the strength of the work published the following year as *Analysis aequationum*, and in 1692 was awarded a Cambridge MA by royal mandate. During 1691 he, Halley, and Newton discussed the publication of some of Newton's papers, but whether Raphson was acquainted with Newton before that date is not known. In the preface to the *Analysis aequationum* he acknowledged Newton's method, as published by Wallis, but believed that his own was different in origin and certainly in procedure.[12] Nevertheless, it gives the same results at each stage as Newton's method, so that Raphson's name has become forever linked to Newton's in the Newton–Raphson method.

Throughout the *Analysis aequationum* Raphson used Harriot's notation, except that he dropped any requirement of dimensional homogeneity. Thus the first type of equation he treated was represented as

$$ba - aaa = c.$$

Raphson then proposed that one should write $a = g + x$ so that we have

$$bg - ggg + \overline{b - 3gg} \times x - 3gxx - xxx = c.$$

If we suppose that g is known (or assumed) then by neglecting terms containing xx or xxx we have an easy approximation for x, given by what Raphson called his *Theorema*:

$$x = \frac{c + ggg - bg}{b - 3gg}. \tag{1}$$

[7]Stirling 1717, 6–18; de Gua 1740, xij–xiij; Cramer 1750, viii–ix. Cramer, for example, wrote: *on découvre que ses principaux guides dans ses Recherches ont été la Doctrine des* Séries infinies, *qui lui doit presque tous, et l'usage du parallelogramme analytique dont il est l'Inventeur.* [one discovers that his principal guide in his research was his doctrine of infinite series, to which he owed almost everything, and the use of the algebraic parallelogram, of which he is the inventor.] Cramer 1750, ix.

[8]See, for example, Stirling 1717, 6–31; Nicole 1738; Maclaurin 1748, 243–273; Maseres 1778

[9]A copy of *Analysis aequationum universalis* inscribed to Wallis by Raphson is in the Bodleian Library, Oxford, Savile G.1, and is digitally accessible through Early English Books Online.

[10]*eximius ille juvenis D. Josephus Ralphson R.S.S.* [that exceptional young man Dr Joseph Raphson FRS.] Halley 1694, 137.

[11]See Thomas 2004.

[12]*sed nec eadem, credo, origine, nec eodem, certe, processu.* Raphson 1690, *Praefatio*.

If we slip into modern notation and write $f(x) = bx - x^3 - c = 0$ for the original equation then we see that Raphson's *Theorema* instructs us to calculate $-\frac{f(g)}{f'(g)}$, as in the modern version of the Newton–Raphson method. Neither Newton nor Raphson, however, used any calculus, but only straightforward algebraic reasoning. Adding the value of x obtained from (1) to the original value of g gives a new value of g, which then becomes the starting point for a new value of x, and so on.[13]

Raphson gave several examples of his method, beginning with the root extraction $aa = 2$ and its corresponding *Theorema*:[14]

$$x = \frac{2 - gg}{2g}.$$

His Problem IX was $aaa - 2a = 5$, which of course was Newton's equation, though Raphson did not say so.[15] For this equation his *Theorema* (similar to (1) above apart from changes in sign) was

$$x = \frac{c + bg - ggg}{3gg - b}.$$

Pursuing the calculation through four iterations Raphson arrived at a solution to 19 decimal places.[16] His results were identical to Newton's for the first two iterations, but his accuracy at the third iteration was better because he retained figures that Newton had rounded off.

Raphson's *Analysis aequationum* was republished several times, but it was not until Lagrange turned to the problem in 1769, almost exactly a century after Newton, that there was any significant new progress.

Lagrange's continued fraction method, 1769 and 1770

Lagrange's work on the solution of numerical equations was presented to the Berlin Academy in three parts, on 20 April and 24 August 1769 and 8 March 1770.[17] The first part, with the title 'Sur la résolution des équations numériques' ('On the solution of numerical equations'), was included in the *Mémoires* of the Academy for the year 1767 (printed in 1769). The second and third parts were published in the volume for 1768 (printed in 1770), under the title 'Additions au mémoire sur la résolution des équations numériques', ('Additions to the memoir on the solution of numerical equations').

[13] *nove autem ista* (x) *per idem Theorema invenitur (mutata, scilicet, semper mutanda* (g) *Ex nova ergo operatione nova rursus enascetur* (g) *& sic ad infinitum.* [moreover a new x is found by the same *Theorema* (changing, obviously, with every change in g). From the new operation therefore there arises in turn a new g, and so on indefinitely.] Raphson 1690, 2.

[14] Raphson 1690, 5.

[15] Raphson 1690, 13

[16] Raphson's solution was 2.09455148154271041 41. The correct solution to 20 decimal places is 2.09455148154232659148, so Raphson's solution was correct only to 11 decimal places. This was an improvement on Newton's solution, however, which was correct only to 7 decimal places.

[17] The presentations are recorded in the Academy *Registre* for 1766–86, BBAW MSI–IV–32, ff. 54v, 60, 71v.

Lagrange, ever aware of his mathematical predecessors, began by noting the methods of Viète and Newton. On Viète's method he had little to say except that it was so long and complicated that it had now been completely abandoned. Newton's method, on the other hand, he described as very simple and easy. He regretted, however, that no-one had paid attention to its drawbacks and imperfections. These were Lagrange's concern here, and he listed four of them.

The first and principal problem of Newton's method, according to Lagrange, was that one is supposed to know the starting value to within a tenth of the correct value; Rolle's method of cascades (see pages 69–70) offers a way of finding approximations to the roots but Lagrange commented that it was not always reliable, especially where the equation has imaginary roots. Second, at each stage one neglects certain terms without knowing their value and therefore without knowing the accuracy of the approximation.[18] Third, one constructs a sequence of approximations which are supposed to converge to the true root, but such convergence may be very slow and indeed there is no guarantee that it will happen at all. Fourth and finally, the method gives only an approximate solution even where there may be a rational solution which can be found exactly; there might be other methods, of course, for finding rational solutions, but Lagrange regarded it as a disadvantage of Newton's method that it does not necessarily discover them.

Lagrange was particularly concerned with the first problem, of finding an appropriate starting value. His idea was to find a quantity Δ less than the difference between any two real roots of the proposed equation. One can then evaluate the polynomial at $0, \Delta, 2\Delta, 3\Delta, \dots$ and any change of sign will indicate the existence of a positive real root in that interval. Further, the smallness of Δ guarantees that all distinct positive roots will be detected. The nearest integer to each will then provide an appropriate starting value for the iteration. The problem is therefore to find a suitable Δ, that is, a lower bound for the differences between the roots.

Suppose, therefore, that the proposed equation is

$$x^m - Ax^{m-1} + Bx^{m-2} - Cx^{m-3} + \cdots = 0,$$

with m roots $\alpha, \beta, \gamma, \dots$. The differences between the roots are the $m(m-1)$ quantities $(\alpha-\beta), (\alpha-\gamma), \dots, (\beta-\alpha), (\beta-\gamma), \dots$. These are therefore the roots of a new equation of degree $m(m-1)$, in u, say, and the required value of Δ is a lower bound for u. Since the differences appear in pairs differing only in sign, this new equation will contain only even powers and may therefore be thought of as an equation of degree $n = \frac{m(m-1)}{2}$ in v where $v = u^2$, namely,

$$v^n - av^{n-1} + bv^{n-2} - cv^{n-3} + \cdots = 0. \qquad (2)$$

Lagrange claimed that the coefficients $a, b, c, \dots$ can be found in terms of A, B, $C, \dots$ using the usual well known properties of the latter. Thus, for example, for a, the

[18]The problem of the value of neglected terms was later one of Lagrange's concerns about power series in general, leading him to his derivation of the Lagrange form of the 'remainder' in Lagrange 1797, §49–§53.

sum of squares of differences, we have

$$\begin{aligned} a &= (\alpha-\beta)^2 + (\alpha-\gamma)^2 + (\beta-\gamma)^2 + \cdots \\ &= (m-1)(\alpha^2+\beta^2+\gamma^2+\cdots) - 2(\alpha\beta+\alpha\gamma+\beta\gamma+\cdots) \\ &= (m-1)(A^2-2B) - 2B \\ &= (m-1)A^2 - 2mB. \end{aligned}$$

Lagrange gave similar but more tedious derivations for $b, c, \ldots$.

If the original equation has pairs of repeated roots, one or more roots of (2) will be zero, and its degree will be correspondingly reduced to $r < n$. Now the substitution $y = \frac{1}{v}$ can be used to transform equation (2) to

$$y^r + \mathfrak{a}y^{r-1} + \mathfrak{b}y^{r-2} + \cdots = 0. \tag{2'}$$

The next stage is to find an upper bound for the roots of (2′), which will be a lower bound for the roots of (2). Lagrange suggested three possible methods. The first was Newton's method, which Lagrange described as the most useful and the most precise (see page 73). The second was to take the absolute value of the greatest negative coefficient and add 1. Lagrange attributed this method to Maclaurin, who had offered a partial proof of it in his *Treatise of algebra* in 1748, but it had first been stated by Rolle in 1690 (see page 69), and proved by Reyneau.[19] Lagrange suggested a third method: suppose $-\mu x^{r-m}$, $-\nu x^{r-n}$, $-\omega x^{r-p}$, … are the negative terms of (2′); then the sum of the two greatest of $\sqrt[m]{\mu}$, $\sqrt[n]{\nu}$, $\sqrt[p]{\omega}$, … will be an upper bound for the roots. Lagrange claimed that this was easy to prove, but did not stop to do it.

Lagrange offered two examples to show how his suggestion worked out in practice. The first was Newton's equation:

$$x^3 - 2x - 5 = 0. \tag{3}$$

Here, the equation for the squares of the differences of the roots is of degree $\frac{3.2}{2} = 3$ and Lagrange found its coefficients, by the rules he had derived earlier, to be −12, 36, 643; thus the equation is

$$v^3 - 12v^2 + 36v + 643 = 0. \tag{4}$$

Lagrange observed that the signs of (4) do not alternate, so it has at least one negative root. This means that (3) has a pair imaginary roots and only one real root. Thus there was no need to seek a minimum distance between the roots; instead one can return to (3) and seek an upper bound for the root directly. By the rule Lagrange had stated earlier, such an upper bound is $\sqrt{2} + \sqrt[3]{5} < 3$. Substituting $x = 0, 1, 2, 3$ in turn into the left hand side of (3) it is easy to see that the root falls between 2 and 3 and that it is closer to 2 than to 3.

[19] Reyneau 1708, 93–96.

Lagrange's second example was the equation

$$x^3 - 7x + 7 = 0. \tag{5}$$

This time the equation for the squares of the differences of the roots is

$$v^3 - 42v^2 + 441v - 49 = 0. \tag{6}$$

Here the alternating signs show that all the roots of (6) are positive, that is, all the roots of (5) are real. Further, the fact that $v = 0$ is not a root of (6) shows that the roots of (5) are distinct. Transforming (6) by putting $y = \frac{1}{v}$ we have

$$y^3 - 9y^2 + \frac{42}{49}y - \frac{1}{49} = 0.$$

Here the absolute value of the largest negative coefficient is 9, and so by Rolle's rule an upper bound for the roots is 10. Newton's rule gives a slightly tighter upper bound, namely, 9. This means that an appropriate lower bound for v is $\frac{1}{9}$, and therefore the required value of Δ is the square root of this, namely, $\Delta = \frac{1}{3}$. Substituting $x = 0, \frac{1}{3}, \frac{2}{3}, \frac{3}{3}, \dots$ into (5) reveals changes of sign between $\frac{4}{3}$ and $\frac{5}{3}$, and also between $\frac{5}{3}$ and $\frac{6}{3}$. This example was presumably chosen by Lagrange to demonstrate that his technique was capable of detecting two distinct roots between the same pair of integers. The negative root of (5) is located by substituting $-x$ for x to give $x^3 - 7x - 7 = 0$, and it is then easy to see that there is a sign change between $x = 3$ and $x = 4$.

One of the remarkable features of Lagrange's procedure is how much previous theory was built into it. Almost all the techniques then known for transforming equation or for discerning the nature of the roots appeared at some point in Lagrange's exposition: Cardano's transformations $x \to x + k$ or $x \to \frac{k}{x}$; Descartes' rule of signs; Rolle's method of cascades; Rolle's rule for an upper bound for the roots; Newton's formulae for sums of powers of the roots; Newton's rule for an upper bound for the roots; and Euler's idea of forming an equation in a function of the roots. Lagrange's work depends on all of these, many of them correctly acknowledged to their original authors.

About half of 'Sur la résolution des équations numériques' is taken up with the problem of locating approximate values of all real and imaginary roots. In the remaining half of the paper Lagrange gave a method for finding those roots precisely, which we will look at only in outline. Suppose we wish to solve the equation

$$Ax^m + Bx^{m-1} + Cx^{m-2} + \cdots + K = 0,$$

and that the preliminary techniques described above show that a root x lies between integers p and $p + 1$. Lagrange's first step was to put $x = p + \frac{1}{y}$, where necessarily $y > 1$. This leads to an equation for y of the form

$$A'y^m + B'y^{m-1} + C'y^{m-2} + \cdots + K' = 0.$$

We now need the largest real value of y (to give the smallest value of $\frac{1}{y}$). Lagrange argued that just as previously one can find the nearest integer below y. Suppose this is q. Then put $y = q + \frac{1}{z}$ and repeat the procedure. This will give the solution as a continued fraction

$$x = p + \cfrac{1}{q + \frac{1}{r + \dots}}. \tag{7}$$

The fraction will terminate if x is rational but can otherwise be continued to give as accurate an approximation as one wishes.

Lagrange observed that the theory of continued fractions had been put to a number of uses but that it had not previously been considered important in connection with equations. He went on to examine the theory of such fractions in detail, proving, for example, that each partial fraction calculated from (7) is closer to the true root than the previous fraction, and indeed closer than any other fraction with a smaller denominator. Further, he was able to give an easily calculated upper bound for the error at any stage of the calculation.

Newton's equation $x^3 - 2x - 5 = 0$ has only one real root so Lagrange's method can be applied without ambiguity, starting from $x = 2 + \frac{1}{y}$, to give

$$x = 2 + \cfrac{1}{10 + \frac{1}{1 + \frac{1}{1 + \frac{1}{2 + \dots}}}}.$$

Thus, according to Lagrange, the partial fractions

$$\frac{2}{1}, \frac{21}{10}, \frac{23}{11}, \frac{44}{21}, \frac{111}{53}, \dots,$$

are alternately smaller or larger than the true value of x. Lagrange was therefore able to deduce that

$$2.09455147 < x < 2.09455149.$$

This was a little more precise than Newton's value of 2.09455147, but less accurate than Raphson's (see page 158 note 16), which Lagrange seems not to have known about.

The second part of Lagrange's treatment, presented in August 1769, four months after the first, offered further suggestions for using the equation for squares of differences, this time for detecting the number of imaginary roots. Here Lagrange found that the number of real roots must belong either to the sequence 1, 4, 5, 8, 9, … or to 2, 3, 6, 7, 10, 11, … (as mentioned above, pages 101–102). This work formed the first part of the 'Additions au mémoire sur la résolution des équations numériques' (§1–§17) .

The third and final part of Lagrange's presentation, in March 1770, was an extended study of continued fractions ('Additions', §18–§67) together with some further refinements to his method of calculating the root ('Additions', §68–§80). These, however, go beyond the scope of our present study.

Chapter 10
The insights of Lagrange, 1771

As described in Chapter 9, Lagrange's paper 'Sur la résolution des équations numériques' and its 'Additions' were presented to the Berlin Academy in the spring and summer of 1769 and in March 1770. Some eighteen months later, in October 1771, Lagrange embarked on another lengthy study of equations: this time investigating algebraic rather than numerical solution. The first three papers of the new set were presented to the Academy in October, November, and December 1771, and the fourth and last in February 1772 after the Christmas break.[1] The first two papers, on cubic and quartic equations respectively, were published as Sections I and II of 'Réflexions sur la résolution algébrique des équations' ('Reflections on the algebraic solution of equations') in the *Nouveaux Mémoires* for 1770 (printed in 1772); the third and fourth papers, on equations of higher degree, appeared as Sections III and IV of the same article, in the *Nouveaux Mémoires* for 1771 (printed in 1773).[2]

Perhaps it was Lagrange's success in numerical solution that encouraged him to turn his mind to the more intractable difficulties of algebraic solution. There was apparently some interest in the subject at the Academy in 1771 because in June of that year Johann Castillon, Astronomer Royal at the Berlin Observatory since 1765, presented a paper entitled 'Mémoire sur les équations résolues par M. de Moivre' ('Memoir on the equations solved by Mr de Moivre'). Castillon engaged in lengthy algebraic manipulations to confirm that the solutions claimed by de Moivre in 1707 were indeed correct (see page 106), but his paper contained nothing new and was unlikely to have inspired Lagrange. It is far more probable that Lagrange was influenced by the appearance in 1768 of Bezout's 'Mémoire sur la résolution générale des équations' (see Chapter 8), a paper that he referred to frequently and in depth in the course of his own work.

Lagrange began by remarking that of all branches of analysis (*l'Analyse*), one might have expected the theory of equations to have reached the greatest degree of perfection, both because of its importance and because of the rapid progress of the earliest discoveries. Indeed, Lagrange believed there was little left to discover on certain topics: the nature of equations; their transformations; conditions for equal roots and a method of finding them; the nature of imaginary roots; rules for discerning whether all the roots of an equation are real and if so how many are positive (or negative). On the other hand, he noted that there was as yet no general rule for finding the number of imaginary roots; or for the number of positive (or negative) roots when the roots

[1]Dates of presentation were 31 October, 28 November, 12 December 1771, and 13 February 1772; see 'Registres de l'Académie depuis le 2 Aoust MDCCLXI [1766] jusq'au 17 Août 1786', Archiv der BBAW, I–IV–32, ff. 101v, 103, 104, 107v.

[2]The first issue of the *Nouveaux Mémoires* replaced the older series of *Mémoires* in 1772.

134 NOUVEAUX MÉMOIRES DE L'ACADÉMIE ROYALE

RÉFLEXIONS

ſur la réſolution algébrique des équations. (*)

PAR M. DE LA GRANGE.

La théorie des équations eſt de toutes les parties de l'Analyſe celle qu'on eût crû devoir aquerir les plus grands degrés de perfection & par ſon importance & par la rapidité des progrès que les premiers inventeurs y ont faits; mais quoique les Géometres qui ſont venus depuis n'ayent ceſſé de s'y appliquer, il s'en faut beaucoup que leurs efforts ayent eu le ſuccès qu'on pouvoit déſirer. On a à la vérité preſque épuiſé tout ce qui concerne la nature des équations, leur transformation, les conditions néceſſaires pour que deux ou pluſieurs racines deviennent égales, ou ayent entr'elles une rélation donnée, & la maniere de trouver ces racines, la forme des racines imaginaires, & la méthode de trouver la valeur de celles qui quoique réelles ſe préſentent ſous une forme imaginaire &c. On a auſſi découvert des regles générales pour reconnoître ſi toutes les racines d'une équation ſont réelles ou non, & pour ſavoir dans le premier cas combien il doit y en avoir de poſitives & de négatives; mais on n'a juſqu'à préſent aucune regle générale pour connoître le nombre des racines imaginaires dans les équations qui doivent en contenir, & moins encore pour ſavoir combien il doit y en avoir de réelles poſitives & de réelles négatives, lorſqu'on connoit d'ailleurs le nombre des réelles & des imaginaires; on n'a pas même une regle pour pouvoir s'aſſurer ſi une équation quelconque propoſée doit contenir quelques racines réelles ou non, à moins que l'équation ne ſoit d'un degré impair, ou que ſon dernier terme ne ſoit négatif.

Ce n'eſt pas qu'on ne puiſſe toujours trouver le nombre des racines imaginaires & des racines réelles poſitives, ou négatives, lorſqu'on connoit la

(*) Ce Mémoire a été lu à l'Académie dans le courant de l'année 1771.

Reflections on the algebraic solution of equations, from Lagrange (1771).

are not all real; nor even for knowing whether an equation has real roots at all unless it is of odd degree. Lagrange claimed that for equations with numerical coefficients all of these matters can be dealt with and that his own methods left little to be desired in this respect; now the problem was to treat the same problems for literal, or general, equations.

It was at this point that Lagrange made the observation from which we began: that with regard to solving such equations there had been scarcely any advance since the time of Cardano. Indeed, he remarked, the first discoveries of the Italian analysts seemed already to have reached the limits of what could be done. All later attempts to push back those limits had succeeded only in producing new methods for solving cubics and quartic equations, but none of those methods seemed applicable to equations of higher degree.

Lagrange therefore proposed to examine the methods in detail, to try to discover exactly why they were not extendable. In doing so he hoped for a double advantage: to shed light on the known solutions for cubic and quartic equations, but also to avoid futile attempts in the search for solutions to equations of higher degree.

Notation. Until now in this book, each author's work has been described as nearly as possible in his own notation. Lagrange, however, in a paper over 200 pages long, changed notation frequently. He never used subscripts (though he did sometimes write $x, x', x'', \ldots$) and so was forced to repeat sections of the alphabet many times, and was by no means consistent in the way he did so. In the account that follows I have standardized notation so that the various arguments in Lagrange's paper can be more easily compared with each other.

For the same reason, I have adopted the following system of marking equations. The first method investigated by Lagrange was the method of Cardano, and in describing this part of his paper I have prefixed all equation numbers by C. Next Lagrange turned to the method proposed by Tschirnhaus; here all equations will be lettered T. Later, he turned to methods suggested by Euler and Bezout; these will be indicated by the letters E and B.

Further, the C, T, E, and B equations in each subsection have been numbered in such a way as to bring out the analogies between the different methods. Thus in each case equation (1) is the original equation in x, with roots we will call $x_1, x_2, x_3, \ldots$; (2) is a substitution in which x is replaced by a new variable y; (3) and (4) are further intermediate equations; (5) is, or yields, a resolvent; (6) gives the solutions $y_1, y_2, \ldots$ of the resolvent; (7) gives the solutions $x_1, x_2, x_3, \ldots$ of the original equation in terms of $y_1, y_2, y_3, \ldots$; while, conversely, equations (8) and (9) give the roots of the resolvent, $y_1, y_2, \ldots$ in terms of the roots of the original equation, $x_1, x_2, x_3, \ldots$.

Cubic equations

The method of Cardano for cubics. Setting aside quadratic equations as both easy and well known, Lagrange began with cubics. This is his account of Cardano's method

(§1–§6). A general cubic, said Lagrange, may be written $x^3 + mx^2 + nx + p = 0$ but since one can always remove the second term one might just as well work (as had del Ferro and Tartaglia) with the simpler form

$$x^3 + nx + p = 0. \tag{C.1}$$

The most natural method of solving such an equation, according to Lagrange, was that suggested by Hudde (see pages 54–55), where one writes

$$x = y + z \tag{C.2}$$

to obtain

$$y^3 + 3y^2z + 3yz^2 + z^3 + n(z + y) + p = 0. \tag{C.3}$$

Lagrange, following Hudde, then separated this into two smaller equations

$$y^3 + z^3 + p = 0 \tag{C.4a}$$

and

$$3yz + n = 0. \tag{C.4b}$$

Lagrange gave no more justification for this step than Hudde had done, but eliminating z from (C.4a) and (C.4b) gave him the by now well known equation for y:

$$y^6 + py^3 - \frac{n^3}{27} = 0. \tag{C.5}$$

Since this is quadratic in y^3 we have

$$y^3 = -\frac{p}{2} \pm \sqrt{q} \tag{C.6}$$

where $q = \dfrac{p^2}{4} + \dfrac{n^3}{27}$. This gives two immediate solutions of (C.5), namely

$$y_1 = \sqrt[3]{-\frac{p}{2} + \sqrt{q}},$$
$$y_2 = \sqrt[3]{-\frac{p}{2} - \sqrt{q}}.$$

The remaining four solutions come from multiplying each of y_1 by α and α^2, where $\alpha^3 = 1$ (but $\alpha \neq 1$). Now we may use equations (C.4b) and (C.2) to find corresponding values of z and x. Since (C.5) yields six possible values for y there are in principle six possible values for x, but it turns out that they are equal in pairs, so that (C.1) has just three solutions:

$$x_1 = \sqrt[3]{-\frac{p}{2} + \sqrt{q}} + \sqrt[3]{-\frac{p}{2} - \sqrt{q}},$$

$$x_2 = \alpha\sqrt[3]{-\frac{p}{2} + \sqrt{q}} + \alpha^2\sqrt[3]{-\frac{p}{2} - \sqrt{q}},$$

$$x_3 = \alpha^2\sqrt[3]{-\frac{p}{2} + \sqrt{q}} + \alpha\sqrt[3]{-\frac{p}{2} - \sqrt{q}},$$

or

$$\begin{aligned} x_1 &= y_1 + y_2, \\ x_2 &= \alpha y_1 + \alpha^2 y_2, \\ x_3 &= \alpha^2 y_1 + \alpha y_2. \end{aligned} \tag{C.7}$$

Like Bezout, Lagrange called equation (C.5) the 'reduced equation' (*la réduite*), but as in Chapter 8, and following Euler, we will call it the resolvent. It is clear from (C.7) that the roots of the original equation depend on the roots of the resolvent. But how, Lagrange asked, do the roots of the resolvent relate in turn to those of the original equation?

To answer this Lagrange returned to the full form of equation (C.1), namely,

$$x^3 + mx^2 + nx + p = 0. \tag{C.1$'$}$$

Since equation (C.1) was obtained from (C.1′) by adding $\frac{m}{3}$ to each root, the roots of (C.1′) are:

$$\begin{aligned} x_1 &= -\frac{m}{3} + y_1 + y_2, \\ x_2 &= -\frac{m}{3} + \alpha y_1 + \alpha^2 y_2, \\ x_3 &= -\frac{m}{3} + \alpha^2 y_1 + \alpha y_2. \end{aligned} \tag{C.7$'$}$$

Multiplying each equation by 1, α or α^2, and using the fact that $1 + \alpha + \alpha^2 = 0$ (because $1 - \alpha^3 = 0$ and $\alpha \neq 1$) it is easy to eliminate m to obtain:

$$\begin{aligned} y_1 &= \frac{x_1 + \alpha^2 x_2 + \alpha x_3}{3}, \\ y_2 &= \frac{x_1 + \alpha x_2 + \alpha^2 x_3}{3}. \end{aligned} \tag{C.8$'$}$$

From this we can see that the six roots of (C.5) correspond to the six permutations of x_1, x_2, x_3; and further that they fall into three pairs y_1, y_2 and αy_1, $\alpha^2 y_2$ and $\alpha^2 y_1$, αy_2 where in each case the product is $-\frac{n}{3}$ as required by (C.4b). If we add to (C.8′) a third equation

$$-\frac{m}{3} = \frac{x_1 + x_2 + x_3}{3}$$

we can solve for x_1, x_2, x_3, to find that each root depends on just one of the conjugate pairs, as seen in (C.7) or (C.7′).

The method of Tschirnhaus for cubics. Lagrange described Tschirnhaus's method in §10–§11 and §15–§16 of his paper. The outline of his argument is as follows. As before, the equation to be solved may be written

$$x^3 + mx^2 + nx + p = 0. \tag{T.1}$$

The transformation suggested by Tschirnhaus was to replace x by a new variable y defined by

$$x^2 = bx + a + y. \tag{T.2}$$

As explained above (pages 58–64), the idea behind this is to choose suitable values of a and b so that the new version of (T.1) will contain neither a square term nor a linear term.

Now, substituting repeatedly for x^2 from (T.2) into (T.1) gives

$$(b^2 + mb + n + a + y)x + (b + m)(a + y) + p = 0$$

or

$$x = -\frac{(b + m)(a + y) + p}{b^2 + mb + n + a + y}. \tag{T.3}$$

Substituting this back into (T.2) then leads to a cubic in y of the form

$$y^3 + Ay^2 + By + C = 0, \tag{T.4}$$

where A and B are polynomials in a, b, m, n, p. Tschirnhaus's method requires that

$$A = B = 0. \tag{T.5}$$

Lagrange's calculations gave him

$$\begin{aligned} A &= 3a - mb - m^2 + 2n, \\ B &= 3a^2 - 2a(mb + m^2 - 2n) + nb^2 + (mn - 3p)b + n^2 - 2mp. \end{aligned}$$

Thus the equation $A = 0$ is of degree 1 in both a and b and the equation $B = 0$ is of degree 2. Equation (T.5) therefore gives rise to two pairs of values of a and b. If either pair is substituted into (T.4) the latter will be reduced to

$$y^3 + C = 0, \tag{T.6}$$

which is easily solved. A single equation derived from (T.5) (obtained by eliminating either a or b) is therefore a resolvent for this method. Note that its coefficients will depend only on m, n, p, the coefficients of the original equation.

Now (T.5) gives rise to two pairs of values of a and b and (T.6) gives three solutions for y (namely, $-\sqrt[3]{C}, -\alpha\sqrt[3]{C}, -\alpha^2\sqrt[3]{C}$). There are thus 2×3 combinations of a, b, y, that can be substituted into (T.3), giving 6 possibilities for x. These turn out to be equal in pairs and therefore reduce to three as one would expect.

It was at this point, as a digression from his work on (T.5), that Lagrange digressed briefly to a discussion of elimination and in particular, the degree of the elimination equation in general (§12–§14; see pages 143–145). An application of his theory predicted, as he had already confirmed by direct calculation, that the resolvent derived from (T.5) must be of degree 2.

Lagrange then showed *a priori* why this must always be the case for Tschirnhaus's method. Since the roots x_1, x_2, x_3 of (T.1) must satisfy (T.2) we have:

$$\begin{aligned} x_1^2 &= bx_1 + a - \sqrt[3]{C}, \\ x_2^2 &= bx_2 + a - \alpha\sqrt[3]{C}, \\ x_3^2 &= bx_3 + a - \alpha^2\sqrt[3]{C}. \end{aligned} \tag{T.7}$$

Eliminating a and $\sqrt[3]{C}$ (by the method also used at (C.7′)) gives

$$b = \frac{x_1^2 + \alpha x_2^2 + \alpha^2 x_3^2}{x_1 + \alpha x_2 + \alpha^2 x_3}. \tag{T.8}$$

In principle, b can take six values as x_1, x_2, x_3 are permuted. It is easy to see, however, multiplying (T.8) by $\frac{1}{1}, \frac{\alpha}{\alpha}, \frac{\alpha^2}{\alpha^2}$ respectively (Lagrange's suggestion), that the six values are equal in threes, so that b takes just two distinct values. Thus the equation for b, which in principle should be of degree six, turns out to be only of degree two.[3]

Further, Lagrange was able to show by algebraic manipulation that the two roots b_1, b_2 of (T.5) are linearly related to y_2^3, y_1^3 obtained from (C.5). The precise relationships are

$$\begin{aligned} b_1 &= \frac{27{y_2}^3 - 2m^3 + 6mn}{3(m^2 - 3n)}, \\ b_2 &= \frac{27{y_1}^3 - 2m^3 + 6mn}{3(m^2 - 3n)}. \end{aligned}$$

Thus, Lagrange claimed, the methods of Cardano and Tschirnhaus are essentially the same.

The methods of Euler and Bezout for cubics. Finally in Section I, Lagrange went on to investigate his third and last method for cubics (§15–§25), that suggested by Bezout in what Lagrange described as 'un excellent Mémoire', the 'Mémoire sur plusieurs classes d'équations' published in 1764 (see pages 115–116). Equations will again be labelled according to the conventions established above, and once again the original equation may be written

$$x^3 + mx^2 + nx + p = 0. \tag{B.1}$$

Lagrange's slightly more general form of the substitution proposed by Bezout in 1764 was

$$x = \frac{f + gy}{k + y}. \tag{B.2}$$

Bezout's idea had been to choose suitable values of f and k (he had supposed $g = 1$) so that the resulting value of y substituted into

$$y^3 + h = 0 \tag{B.6}$$

[3]Strictly speaking it is the cube of an equation of degree 2, since each pair of roots is repeated three times.

would yield (B.1). Lagrange noted that the required values of f, g, h, and k are thus to be found by elimination of y from (B.2) and (B.6). He also saw immediately that the three possible values of y from (B.6) are $-\sqrt[3]{h}$, $-\alpha\sqrt[3]{h}$, and $-\alpha^2\sqrt[3]{h}$ and that these in turn will give rise to the three roots of (B.1) (it had taken Bezout rather longer to arrive at the same conclusion).

Further, Lagrange saw that (B.2) can be combined with (B.6) as follows:

$$\begin{aligned} x &= \frac{f+gy}{k+y} \\ &= \frac{(f+gy)(k^2-ky+y^2)}{k^3+y^3} \\ &= \frac{(k^2f-hg)+(k^2g-kf)y+(f-kg)y^2}{k^3-h}. \end{aligned}$$

In other words, x is of the form $a+by+cy^2$. Thus the form

$$x = a + by + cy^2 \tag{BE.2}$$

suggested by Euler in 'De resolutione aequationum cuiusvis gradus' (written in 1759, published in 1764; see Chapter 5) and the transformation proposed by Bezout in his 'Mémoire sur la résolution générale des équations' (written by 1763, published in 1768; see Chapter 8) are both equivalent to (B.2). Lagrange commented that the only difference between Bezout's method and Euler's was that Bezout took $h = -1$ whereas Euler had allowed any one of a, b, c, or h to be 1, as convenient.

Using Bezout's value of $h = -1$ the three solutions given by (BE.2) are

$$\begin{aligned} x_1 &= a+b+c, \\ x_2 &= a+b\alpha+c\alpha^2, \\ x_3 &= a+b\alpha^2+c\alpha, \end{aligned} \tag{B.7}$$

from which we see, eliminating a and either b or c in the usual way, that

$$\begin{aligned} b &= \frac{x_1+\alpha^2x_2+\alpha x_3}{3}, \\ c &= \frac{x_1+\alpha x_2+\alpha^2x_3}{3}. \end{aligned} \tag{B.8}$$

Thus b and c correspond precisely to y_1 and y_2 in Cardano's method (C.8′). As a pair they take three sets of values as x_1, x_2, x_3 are permuted. The details of Euler's method vary according to which of a, b or c is taken to be 1, but essentially differ little from Bezout's.

From these investigations Lagrange concluded that the methods of Cardano, Tschirnhaus, Bezout, or Euler, though they differ in detail, have some striking features in common:

(i) In each method one arrives at a resolvent equation, whose roots determine the roots of the original equation.

(ii) For all the methods the roots of the resolvent consist of multiples of either

$$y = x_1 + \alpha^2 x_2 + \alpha x_3$$

or

$$y = (x_1 + \alpha^2 x_2 + \alpha x_3)^3.$$

In the first case, y can take 6 possible values as x_1, x_2, x_3 are permuted so the resolvent will be of degree 6, but will contain only third and sixth powers of y and will therefore be solvable as an equation of degree 2 (as in the method of Cardano and the second method of Bezout). In the second case, y can take only 2 values and the resolvent will therefore immediately be of degree 2 (as in the methods of Tschirnhaus and Euler and the first method of Bezout). Thus whatever method is chosen, a cubic equation can be solved by means of a resolvent of degree 2.

Quartic equations

Up to the end of the seventeenth century there were essentially just two methods for solving quartic equations, that of Ferrari and that of Descartes. Tschirnhaus had proposed a third method but had not worked out the details. In the eighteenth century the methods of Euler and Bezout were also shown to be applicable to quartics.

The method of Ferrari for quartics. In describing Lagrange's treatment of the method of Ferrari (§26–§30) we will keep to the notation introduced above for the roots of the original equation (x_1, x_2, ...) and supplementary variables (y, z). Equations will be lettered F with the same numbering conventions as above.

As usual we may suppose that the second term of the equation has been removed so that the proposed quartic may be written as

$$x^4 + nx^2 + px + q = 0. \qquad \text{(F.1)}$$

Ferrari's technique was to introduce a second unknown, here called y, so that x^4 is replaced by $(x^2 + y)^2$ and (F.1) becomes

$$(x^2 + y)^2 = (2y - n)x^2 - px - q + y^2. \qquad \text{(F.2)}$$

Clearly the left-hand side of (F.2) is a perfect square. For the right-hand side to be so we require

$$(2y - n)(y^2 - q) - \frac{p^2}{4} = 0,$$

that is,

$$y^3 - \frac{n}{2}y^2 - qy + \frac{4nq - p^2}{8} = 0, \qquad \text{(F.5)}$$

which is the resolvent for this method. If any solution of (F.5) is substituted into (F.2), that equation takes the form

$$(x^2 + y)^2 = (2y - n)\left(x - \frac{p}{2(2y - n)}\right)^2,$$

or

$$(x^2 + y)^2 = z^2\left(x - \frac{p}{2z^2}\right)^2$$

where $z^2 = 2y - n$. Taking square roots of both sides we arrive at two quadratic equations in x, namely,

$$x^2 \mp zx + y \pm \frac{p}{2z} = 0 \tag{F.3}$$

whose four solutions are

$$\begin{aligned} &\frac{z}{2} + \sqrt{\frac{z^2}{4} - \frac{p}{2z} - y},\\ &\frac{z}{2} - \sqrt{\frac{z^2}{4} - \frac{p}{2z} - y},\\ &-\frac{z}{2} + \sqrt{\frac{z^2}{4} + \frac{p}{2z} - y},\\ &-\frac{z}{2} - \sqrt{\frac{z^2}{4} + \frac{p}{2z} - y}. \end{aligned} \tag{F.4}$$

From (F.3) it is easy to see that if x_1, x_2, x_3, x_4 are the four roots of the original equation then there are pairs x_1, x_2 and x_3, x_4, such that

$$x_1 + x_2 = z,$$
$$x_3 + x_4 = -z,$$

and

$$x_1x_2 = y + \frac{p}{2z},$$

$$x_3x_4 = y - \frac{p}{2z}.$$

From these we obtain

$$y = \frac{x_1x_2 + x_3x_4}{2} \tag{F.8a}$$

and

$$z = \frac{(x_1 + x_2) - (x_3 + x_4)}{2}. \tag{F.8b}$$

Under permutations of the roots x_1, x_2, x_3, x_4, we see that y can take 3 values and is therefore necessarily the root of an equation of degree 3 (namely, (F.5)), while z can take 6 values (in three pairs of opposite sign).

The method of Descartes for quartics. Lagrange passed straight from the method of Ferrari to that of Descartes (§33–§37). As before, we may assume that the equation to be solved is

$$x^4 + nx^2 + px + q = 0. \tag{D.1}$$

Descartes' method requires (D.1) to be written as the product of two quadratic factors. In other words we need to find coefficients f, g, k such that

$$x^4 + nx^2 + px + q = (x^2 - fx + g)(x^2 + fx + k). \tag{D.3}$$

Multiplying out, equating coefficients, and eliminating g and k leads to the resolvent

$$f^6 - 2nf^4 + (n^2 - 4q)f^2 - p^2 = 0. \tag{D.5}$$

It is clear from (D.3) that if x_1, x_2, x_3, x_4 are the roots of (D.1) there are pairs x_1, x_2 and x_3, x_4 such that

$$x_1 + x_2 = f,$$
$$x_3 + x_4 = -f.$$

Thus f here is equivalent to z in Ferrari's method; indeed the two equations (F.3) give rise to the two factors in (D.3) and vice versa. Thus the two methods are equivalent.

The method of Tschirnhaus for quartics. Lagrange completed his Section II on quartic equations with the methods of Tschirnhaus (§38–§45), and Euler and Bezout (§46–§49).

In extending the method of Tschirnhaus to quartics Lagrange began with a general quartic equation with a full complement of terms

$$x^4 + mx^3 + nx^2 + px + q = 0. \tag{T.1}$$

Lagrange saw that in fact he only needed to transform (T.1) into a quadratic equation (in y^2, say), so that only two terms need be eliminated. He therefore proposed a substitution of the form

$$x^2 + fx + g + y = 0. \tag{T.2}$$

The somewhat lengthy process of eliminating x from (T.1) and (T.2) yields a quartic of the form

$$y^4 + Ay^3 + By^2 + Cy + D = 0, \tag{T.4}$$

where A, B, C, D, are polynomials in f, g, m, n, p, q. To reduce this to a quadratic equation in y^2, namely

$$y^4 + By^2 + D = 0, \tag{T.6}$$

it is sufficient to discover the conditions under which

$$A = 0 \quad \text{and} \quad C = 0. \tag{T.5}$$

Lagrange's calculations showed that the equation $A = 0$ is linear in f and g but the equation $C = 0$ is cubic in f and g, thus leading to three pairs of values of f and g. Once any pair is found it is possible to reduce (T.4) to (T.6) and then to solve (T.6) for y and (T.2) for x.

Suppose that the solutions of (T.6) are $\pm y_1$, $\pm y_2$. Then from (T.2) we have

$$\begin{aligned} x_1^2 + fx_1 + g + y_1 &= 0, \\ x_2^2 + fx_2 + g - y_1 &= 0, \\ x_3^2 + fx_3 + g + y_2 &= 0, \\ x_4^2 + fx_4 + g - y_2 &= 0, \end{aligned} \tag{T.7}$$

and eliminating y_1, y_2, and g from (T.7) yields

$$f = -\frac{(x_1^2 + x_2^2) - (x_3^2 + x_4^2)}{(x_1 + x_2) - (x_3 + x_4)}. \tag{T.8}$$

Clearly f takes only three values as the roots x_1, x_2, x_3, x_4 are permuted, which explains why the equation for f, obtained by eliminating g from $A = 0$ and $C = 0$ at (T.5), must be cubic.

The substitutions of Euler and Bezout for quartics. Finally, Lagrange applied the substitutions suggested by Euler and Bezout to the same general equation

$$x^4 + mx^3 + nx^2 + px + q = 0. \tag{EB.1}$$

This time the required substitution is

$$x = a + by + cy^2 + dy^3, \tag{EB.2}$$

with the additional condition

$$y^4 + D = 0. \tag{EB.6}$$

Here, there are five unknown quantities a, b, c, d, and D but only four equations from (EB.2), and so one of the quantities may be arbitrarily chosen. Euler in 1759 had worked with $c = 1$ and arrived at an equation in D; while Bezout by 1763 had stipulated that $D = -1$ and so arrived at a resolvent in c. Either way there are four solutions for y, namely $\pm k$, $\pm\sqrt{-1}k$ where $k^4 = -D$. From (EB.2) the solutions of the original equation are then

$$\begin{aligned} x_1 &= a + bk + ck^2 + dk^3, \\ x_2 &= a - bk + ck^2 - dk^3, \\ x_3 &= a + \sqrt{-1}bk - ck^2 + \sqrt{-1}dk^3, \\ x_4 &= a - \sqrt{-1}bk - ck^2 + \sqrt{-1}dk^3. \end{aligned} \tag{EB.7}$$

From Euler's calculation, with $c = 1$ we arrive at

$$D = -\frac{((x_1 + x_2) - (x_3 + x_4))^2}{16}, \tag{E.8}$$

so that D is simply a multiple of the square of z from equation (F.8b) or of f from (D.5). Using Bezout's method, on the other hand, with $D = -1$, we find that

$$c = \frac{(x_1 + x_2) - (x_3 + x_4)}{4} \tag{B.8a}$$

and

$$b = (x_1 - x_2) - \sqrt{-1}(x_3 - x_4). \tag{B.8b}$$

Thus c from (B.8a) is the same as $\frac{1}{2}z$ from (F.8b). Under permutations of the roots it takes six values, in three pairs of opposite sign, so the resolvent in c will be a cubic equation in c^2. In principle b can take 24 values. However, as Lagrange pointed out, swapping x_1 for x_2, and x_3 for x_4, simply transforms b to $-b$. Other exchanges transform b to $\sqrt{-1}b$ or to $-\sqrt{-1}b$, but in all cases b^4 remains the same. The twenty-four values of b therefore give rise to only six values of b^4, so the resolvent in b will be an equation of degree six in b^4.

Further, Lagrange noted that b in (B.8b) combines each of x_1, x_2, x_3, and x_4 in turn with distinct fourth roots of 1, just as y_1 and y_2 in (C.8′) (Cardano's method for cubics) combine each of x_1, x_2, and x_3 with distinct cube roots of 1. Thus he was able to argue that there was a fundamental similarity between the structure of the results for quartic and cubic equations (as Euler noted in 1753, see page 110). This was something he went on to explore much further in his Section IV.

Lagrange offered very much more detail than has been presented here, exploring the relationships between various quantities at considerable length. The foregoing should be enough, however, to justify the main conclusions that Lagrange himself came to in the final paragraph of Section II. In all cases solving the original quartic (always labelled (1) above) depends upon being able to solve a reduced or resolvent equation (labelled (5)), sometimes of degree 3, sometimes of degree 6 but reducible to 3. Further, the roots of the resolvent are always functions of the roots x_1, x_2, x_3, x_4 of the original equation, and such functions take only a limited number of values as the roots are permuted. The function $x_1x_2 + x_3x_4$ from (F.8a), for example, takes three values; while the function $(x_1 + x_2) - (x_3 + x_4)$ from (F.8b), (E.8), (B.8a), takes six, in three pairs with opposite sign. Lagrange had also found other and more complicated functions with six values, falling into three pairs each with the same sums and products. In the closing words of Section II, Lagrange stated a new and far-reaching conclusion: that the solution of quartic equations depends solely upon the existence of such functions.[4]

---

[4] *C'est uniquement de l'existence de telles fonctions que dépend la résolution générale des équations du quatrieme degré.* Lagrange 1770, 215.

Higher degree equations. In Section III of his paper (§51–§85), Lagrange turned his attention to equations of degree five or more. Here, he said, he knew of only two methods with any hope of success: that of Tschirnhaus, and that of Euler and Bezout. These had been shown to work for cubic and quartic equations, but it was already clear that the corresponding calculations for quintic equations were very difficult. Lagrange could see that the method of Tschirnhaus, for instance, would lead to a resolvent of degree 24. Euler by his method had arrived at the same conclusion but had hoped, by analogy with cubic and quartic equations, that the resolvent would reduce to degree 4; Bezout, however, had shown that there was no reason to suppose that this was so (see Chapter 8). Indeed, Bezout had argued that a resolvent for a quintic equation would in general be of degree 120, but might contain only powers that are multiples of five, and would therefore reduce to degree 24. Bezout thought that the solution to such an equation would involve only fourth or lower roots, and therefore in principle be no more difficult than a quartic. Lagrange, however, could see no reason for this to be true.

The outcome of these 'réflexions' was that Lagrange doubted that any of the methods so far described could offer a complete solution for equations of degree 5, even less for equations of higher degree.[5] This uncertainty, combined with the length of the calculations, was enough to deter anyone from even trying to resolve what he described as one of the most famous and important problems of algebra.[6] It was therefore all the more important to try to judge in advance what success could be hoped for, and so Lagrange proposed to carry out the same kind of analysis of higher degree equations as he had for cubics and quartics.

Here we will give an outline of his findings, using the same conventions of notation and labelling as previously.

Suppose the proposed equation of degree μ is

$$x^{\mu} + mx^{\mu-1} + nx^{\mu-2} + \cdots = 0, \tag{1}$$

and that we make a substitution of the form suggested by Tschirnhaus,

$$x^{\mu-1} + fx^{\mu-2} + gx^{\mu-3} + \cdots + l + y = 0, \tag{2}$$

leading to the transformed equation

$$y^{\mu} + Ay^{\mu-1} + By^{\mu-2} + \cdots = 0. \tag{3}$$

Ideally we now want

$$A = B = C = \cdots = 0 \tag{5}$$

[5] *Il résulte de ces réflexions qu'il est très douteux que les méthodes dont nous venons de parler puissent donner la résolution complette des équations du cinquieme degré, & à plus forte raison celle des degrés supérieurs;* Lagrange (1771) [1773], 140.

[6] *un des problemes les plus célebres & les plus importans de l'Algebre.* Lagrange (1771) [1773], 140 (with original spellings).

so that (3) reduces to the simple form

$$y^{\mu} + V = 0. \tag{6}$$

This has solutions

$$\begin{aligned} y_1 &= u, \\ y_2 &= \alpha u, \\ y_3 &= \alpha^2 u, \\ &\dots \\ y_{\mu-1} &= \alpha^{\mu-1} u, \end{aligned}$$

where $\alpha^{\mu} = 1$ (with $\alpha \neq 1$) and u is some fixed value such that $u^{\mu} = -V$. Each of these solutions gives rise to a corresponding solution of (2). The latter therefore satisfy

$$\begin{aligned} x_1^{\mu-1} + f x_1^{\mu-2} + g x_1^{\mu-3} + \dots + l + u &= 0, \\ x_2^{\mu-1} + f x_2^{\mu-2} + g x_2^{\mu-3} + \dots + l + \alpha u &= 0, \\ x_3^{\mu-1} + f x_3^{\mu-2} + g x_3^{\mu-3} + \dots + l + \alpha^2 u &= 0, \\ &\dots \\ x_{\mu}^{\mu-1} + f x_{\mu}^{\mu-2} + g x_{\mu}^{\mu-3} + \dots + l + \alpha^{\mu-1} u &= 0. \end{aligned} \tag{7}$$

From these μ equations it is in principle possible to determine the μ quantities f, g, ..., l, and u, in terms of x_1, x_2, ..., and α.

Now, since f, for instance, can take $\mu!$ values as x_1, x_2, ..., x_{μ} are permuted, it must satisfy an equation of degree $\mu!$. However, any one of u, αu, $\alpha^2 u$, ... could have been chosen as y_1 in the first equation of (7), with the rest following in order. Or, equivalently, any of the roots x_i could have been chosen as x_1 with the rest following in order. That is, what we now call a cyclic permutation of $x_1, x_2, \dots, x_{\mu}$ can make no difference to the value of f. Therefore, the degree $\mu!$ of the equation for f must reduce to $(\mu-1)!$ or, more accurately, the equation of degree $\mu!$ for f must be reducible, with μ factors each of degree $(\mu-1)!$. This argument confirmed for Lagrange the results he had already discovered by direct calculation for $\mu = 2, 3, 4$, or 5.

When μ is composite it may be possible to find a resolvent of degree even lower than $(\mu-1)!$ If $\mu = \nu\pi$, for example, then instead of requiring $A = B = C = \dots = 0$ in equation (5) we may simply wish to find an equation of degree ν in powers of y^{π}, as Lagrange had done for quartic equations in applying the method of Tschirnhaus (see (T.6) above). In this case one needs to eliminate only $\nu(\pi-1)$ unknowns, and Lagrange claimed that the degree of the resolvent will then be only

$$\frac{(\mu-1)(\mu-2)\dots(\nu+2)(\nu+1)\nu}{\pi^{\nu-1}}.$$

Thus in applying the method of Tschirnhaus to quartics, where $\mu = 4$, $\nu = 2$, $\pi = 2$, the degree of the resolvent is

$$\frac{3.2}{2} = 3.$$

To discover whether any further reduction was possible in general, Lagrange first considered the case where μ is prime (§56, §57). Here he argued that not only was the choice of u in (7) arbitrary, but also the choice of α in the solutions of (6): any μ^{th} root of 1 (except 1 itself) serves the same purpose, and there are $\mu - 1$ of them. Each equation of degree $(\mu-1)!$ therefore reduces further, to $\mu-1$ factors of degree $(\mu-2)!$. Similar but more complicated arguments apply to the case where μ is composite (§59–§64), where now the degree of a resolvent is seen to depend upon the number and multiplicity of prime factors of μ.

Finally (§69–§85), Lagrange explored the properties of a particular function of the roots that had by now appeared many times, both in his own work and in the papers of Euler and Bezout, namely,

$$t = x_1 + \alpha x_2 + \alpha^2 x_3 + \cdots + \alpha^{\mu-1} x_\mu.$$

Clearly t can take $\mu!$ values as $x_1, x_2, \ldots, x_\mu$ are permuted. However, if t_1 arises as a particular value, so will $\alpha t_1, \alpha^2 t_1, \ldots, \alpha^{\mu-1} t_1$, all of which satisfy $t^\mu = \theta$ for some value of θ. Thus, θ can take at most $(\mu-1)!$ values.

Now if μ is prime it is easy to see that θ, given by

$$(x_1 + \alpha x_2 + \alpha^2 x_3 + \cdots + \alpha^{\mu-1} x_\mu)^\mu,$$

takes $\mu - 1$ values $\theta_1, \theta_2, \ldots, \theta_{\mu-1}$ as α is replaced by $\alpha^2, \alpha^3, \ldots, \alpha^{\mu-1}$. These values of θ are therefore the roots of an equation of the form

$$\theta^{\mu-1} - T\theta^{\mu-2} + U\theta^{\mu-3} - \cdots = 0.$$

The equation of degree $(\mu-1)!$ for θ therefore decomposes into $(\mu-2)!$ factors of degree $\mu - 1$, that is, the coefficients $T, U, \ldots$ can take $(\mu-2)!$ sets of values.

All this is easily illustrated for the case $\mu = 3$. Suppose the three roots of the original equation

$$x^3 + px + q = 0$$

are x_1, x_2, x_3, and let

$$t = x_1 + \alpha x_2 + \alpha^2 x_3.$$

where $\alpha^3 = 1$ and $\alpha \neq 1$. Clearly t can take six possible values as x_1, x_2, x_3 are permuted. If we fix

$$t_1 = x_1 + \alpha x_2 + \alpha^2 x_3$$

then successive applications of the cyclic permutation (x_1, x_2, x_3) generate $t_2 = \alpha^2 t_1$ and $t_3 = \alpha t_1$, with the property $t_1^3 = t_2^3 = t_3^3 = \theta_1$. Replacing α by α^2 throughout gives the three remaining possibilities

$$t_4 = x_1 + \alpha^2 x_2 + \alpha x_3$$

and $t_5 = \alpha^2 t_4$ and $t_6 = \alpha t_4$, with the property $t_4^3 = t_5^3 = t_6^3 = \theta_2$. Thus θ_1 and θ_2 are the roots of an equation of degree 2! of the form

$$\theta^2 - T\theta + U = 0$$

where $T = \theta_1 + \theta_2$ and $U = \theta_1\theta_2$. It is a little tedious but not intrinsically difficult to work out that $T = 27q$ and $U = -27p^3$ (as Lagrange had long ago established for Cardano's method) so that the equation for θ is

$$\theta^2 - 27q\theta - 27p^3 = 0.$$

Thus for a cubic, the equation for θ is of degree 2!, which can be considered to 'decompose' into 1! factor of degree 2 of the form $\theta^2 - T\theta + U = 0$; and T and U can be expressed in terms of the coefficients of the original equation.

Lagrange made a lengthy attempt to extend the same reasoning to the case where $\mu = 5$. Now the equation for θ is of degree 4! with 3! factors of degree 4, each of the form

$$\theta^4 - T\theta^3 + U\theta^2 - X\theta + Y = 0.$$

That is to say, the equations for each of T, U, X, Y are of degree 3!, but Lagrange could find no way of reducing them to lower degree. Thus, as so often with quintic equations, the attempt to solve them led in practice only to greater difficulties than one had started with.

In the remainder of Section III (§75–§84) Lagrange considered the function t in the case where μ is composite but again achieved only partial results.

Lagrange's theorem

Section IV of Lagrange's paper (§86–§115), presented to the Berlin Academy in February 1772, built on his findings in the first three sections but now Lagrange began to move away from the initial problem of solving equations to a more general examination of properties of functions of their roots. In his opening paragraph he summarized his findings so far by claiming that solving equations always comes down to the same general principle, namely, finding a function of the roots with two crucial properties: (1) that it satisfies a reduced or resolvent equation with degree lower than that of the original; (2) that the roots of the original equation can be easily recovered from it. The art of solving equations, then, is to discover such functions. Whether they even exist, however, for equations of a given degree was a question to which there was as yet no general answer.

Lagrange introduced the notation $f : (x')(x'')(x''') \ldots$ for a general function of the roots $x', x'', x''', \ldots$ of an equation, with the convention that $f : (x', x'')(x''') \ldots$, for example, denotes a function that is not changed by transposition of the first two variables.[7] In Lagrange's terminology, the function keeps the same 'value' (*valeur*) when the variables in the first two places are transposed.

In the remainder of this discussion we will adopt, as before, the more easily handled subscript notation $x_1, x_2, x_3, \ldots, x_\mu$ for the roots of an equation of degree μ. A function f of these roots can take $\mu!$ values as the roots are permuted, and so, Lagrange argued,

[7] Such a function might be, for instance, $x'x'' + x'''$ or $(x' + x'')x'''$.

these values, denoted in general by t, must be the roots of an equation of order $\pi = \mu!$, which he wrote as

$$\Theta = t^{\pi} - Mt^{\pi-1} + Nt^{\pi-2} - Pt^{\pi-3} + \cdots = 0.$$

He also claimed that $M, N, \ldots$ are functions of the coefficients $m, n, p, \ldots$ of the original equation. He gave some theoretical justification for this for equations of degree 2, 3, or 4 (in §90–§96), but he also referred to the results he had found by direct calculation on several occasions for cubic and quartic equations.[8]

Lagrange had shown in his earlier examples based on cubic and quartic equations that the degree of the equation $\Theta = 0$ is reduced in cases where f remains unchanged under certain permutations of the roots. He now explained more generally how this could occur (§97). Before looking at his own example, however, which was not entirely straightforward, we will consider a rather easier one.

Suppose we have a function $f : (x_1\, x_2\, x_3)(x_4) \ldots$ invariant under any of the six permutations of the variables in the first three places.[9] It is clear that however the roots are permuted (or labelled) the values of f will be equal in sets of six. That is, the degree of the equation $\Theta = 0$ for the values of f reduces from $\mu!$ to $\frac{\mu!}{3!}$.

Now let us return to Lagrange, who considered a function of the form

$$f : (x_1)(x_2)(x_3)(x_4) \ldots .$$

He next supposed that

$$f : (x_1)(x_2)(x_3)(x_4) \ldots = f : (x_2)(x_3)(x_1)(x_4) \ldots .$$

Lagrange described this by saying that the function keeps the same value when we change x_1 to x_2, x_2 to x_3, and x_3 to x_1. What he meant was that the function will remain unchanged under a cyclic permutation of the first three variables, whichever three happen to be chosen.[10] This is clear from the next few lines of his discussion where he claimed that it must also be the case that

$$f : (x_4)(x_3)(x_1)(x_2) \ldots = f : (x_3)(x_1)(x_4)(x_2) \ldots .$$

At this point Lagrange failed to see the full implications of his argument, for in both cases the same reasoning should have given him a third equal value, that is,

---

[8] *Et comme nous avons démontré ci-dessus que l'expression de* Θ *doit être nécessairement une fonction rationelle de t & des coëfficiens m, n, p &c. de l'équation proposée; il s'ensuit que les quantités* M, N, P *&c. seront nécessairement des fonctions rationelles de m, n, p &c. qu'on pourra trouver directement, comme nous l'avons pratiqué dans les Sections précédentes.* [And as we have demonstrated above that the expression for Θ must necessarily be a rational function of t and the coefficients m, n, p, etc. of the proposed equation, it follows that the quantities M, N, P, etc. will necessarily be rational functions of m, n, p, etc. which one could find directly, as we have done in the preceding Sections.] Lagrange (1771) [1773] §96.

[9] For example, $x_1 x_2 x_3 + x_4$ or $x_1 + x_2 + x_3 + x_4^2$.

[10] A function of this type could be, for example, $x_1^2 x_2 + x_2^2 x_3 + x_3^2 x_1 + x_4$.

$$f : (x_1)(x_2)(x_3)(x_4)\ldots = f : (x_2)(x_3)(x_1)(x_4)\ldots = f : (x_3)(x_1)(x_2)(x_4)\ldots$$

and

$$f : (x_4)(x_3)(x_1)(x_2)\ldots = f : (x_3)(x_1)(x_4)(x_2)\ldots = f : (x_1)(x_4)(x_3)(x_2)\ldots.$$

Lagrange missed this point and claimed only that we will have values of f that are equal in pairs. Thus, he claimed, the roots of $\Theta = 0$ must be equal in pairs and so Θ is equal to a square θ^2, and the equation $\Theta = 0$ is reduced to the equation $\theta = 0$ with degree $\frac{\mu!}{2}$. Since the roots of $\theta = 0$ are actually repeated in threes, what he should have said was that Θ is equal to a cube θ^3, and the equation $\Theta = 0$ is therefore reduced to the equation $\theta = 0$ with degree $\frac{\mu!}{3}$.

Lagrange's argument was wrong in its details, but his insight was essentially right. In any case he quickly corrected himself, for in §98 he asserted that a function of the form

$$f : (x_1 x_2 x_3)(x_4)\ldots$$

will satisfy an equation of degree $\frac{\mu!}{3!}$ (as argued above), while a function of the form

$$f : (x_1 x_2)(x_3 x_4)(x_5)\ldots$$

will satisfy an equation of degree $\frac{\mu!}{2!2!}$. And in general a function of the form

$$f : (x_1 x_2 \ldots x_\alpha)(x_{\alpha+1} \ldots x_{\alpha+\beta})(x_{\alpha+\beta+1} \ldots x_{\alpha+\beta+\gamma})\ldots$$

will satisfy an equation of degree $\frac{\mu!}{\alpha!\beta!\gamma!\ldots}$.

Another way of looking at the above theorem is that the number of values of a function of μ variables must divide $\mu!$. For almost a century this theorem was known as *le théorème de Lagrange* or Lagrange's Theorem. Later, a related theorem came to acquire the same name in the quite different context of group theory. In modern terms, the set of permutations that leave the values of f unchanged is a subgroup S of the group S_μ of the $\mu!$ permutations of μ variables (S is now known as the stabiliser). The number of *different* values of f is the index of S in S_μ, namely, $\frac{|S_\mu|}{|S|}$. What Lagrange had demonstrated was that the *index* divides $|S_\mu|$, but this can equally be interpreted as saying that $|S|$ itself divides $|S_\mu|$. A much more general version of this theorem, now also known as Lagrange's Theorem, is that the order of any subgroup of a finite group divides the order of that group.

The next part of Section IV (§100–§104) is taken up with a theorem that Lagrange claimed as one of the most important in the theory of equations. Suppose we have two functions t and y of the roots $x_1, x_2, \ldots, x_\mu$. In his initial statement of the theorem in §100, Lagrange claimed that given a value of t one could find a corresponding value of y.[11] Some care is needed in interpreting this statement (which Lagrange himself expanded upon over several pages).

[11] *Or, dès qu'on aura trouvé, soit par la résolution de léquation $\theta = 0$, ou autrement, la valeur d'une fonction donnée des racines x', x'', x''' &c. je dis qu'on pourra trouver aussi la valeur d'une autre fonction quelconque des mêmes racines, & cela, généralement parlant, par le moyen d'un équation simplement*

In the first place Lagrange restricted himself to functions that he had defined earlier (in §88) as 'similar' (*semblables*), in which every permutation of the roots that changes t also changes y and *vice versa.* That is, the number of different values of t is the same as the number of different values of y. Lagrange proved that in this case each value of y may be written as a rational function in the values of t (or *vice versa*).[12] An example of such a pair of functions (denoted by y and b) and of the relationships between their values can be seen above on page 169 at the end of the discussion of Tschirnhaus's method for cubics.

The most useful application of the theorem is to regard the roots themselves, x_1, x_2, x_3, … as the values of the function y. As Lagrange put it: take the root x in place of the function y (or t) and apply the preceding conclusions.[13]

Applying his results to the case $\mu = 3$ (§105, §106) led Lagrange to examine in detail the possible permutations of three roots and the nature of functions f that remained invariant under them. His investigations confirmed what he had discovered long ago (in §5), that for the solution of a cubic equation the required function in its simplest form is $A(x_1 + \alpha^2 x_2 + \alpha x_3)$ where $\alpha^3 = 1$ and A is a constant. In the case $\mu = 4$ (§107, §108) he discovered, again by careful examination of the possible permutations, that suitable candidates for f are (1) $(x_1+x_3)-(x_2+x_4)$ or $x_1x_3-x_2x_4$ (as he had already found in §31, §32); or else (2) $x_1 + \alpha x_2 + \alpha^2 x_3 + \alpha^3 x_4$ (as he had found in §47). All of this thus served to confirm what Lagrange had discovered by direct investigation in Sections I and II, causing him to repeat yet again that the problem of solving equations led one to a calculus of permutations (§109).[14] The application of similar principles to quintics or equations of higher degree, said Lagrange, was clearly going to require a good deal of further research, to which he hoped to return in the future. For now though, he concluded, he was satisfied to have put in place the foundations of what seemed to him a new and general theory.

Lagrange was correct in his perception of what he had achieved. The search for algebraic solutions to quintics or equations of higher degree was not over, but Lagrange's work suggested quite strongly that such solutions might in general be impossible to find, as was later proved to be the case (see Chapter 11). More crucially, however, Lagrange had shifted the entire discourse on equation-solving away from a hunt for effective techniques towards an examination of the fundamentals of the problem. There

---

linéaire, à l'exception de quelques cas particulers qui exigent une équation du second degré ou du troisieme &c. [Now, as soon as one has found a value of a given function of the roots x_1, x_2, x_3, … (whether by solving the equation $\theta = 0$ or otherwise) I say that one can also find a value of any other function of the same roots, generally speaking by means of an equation that is simply linear, except for some particular cases which require and equation of second or third degree.] Lagrange (1771) [1773] §100.

[12]The situation is more complicated if the condition of similarity does not hold, but Lagrange was able to give some partial results in (1771) [1773] §103, §104.

[13]*il n'y aura pour cela qu'à prendre la simple racine* **a** *à la place de la fonction* y, *$ appliquer à ce cas les conclusions précédentes.* Lagrange (1771) [1773] §104.

[14]*Voilà, si je ne me trompe, les vrais principes de la résolution des équations, & l'analyse la plus propre á y conduire; tout se réduit, comme l'on voit, à une espece de calcul des combinations, par lequel on trouve* a priori *les résultats auxquels on doit s'attendre.* [Here, if I am not mistaken, are the true principles of solving equations, and the most correct analysis to lead there; all of which reduces, as one sees, to a kind of calculus of combinations, by which one finds *a priori* the results one must expect.] Lagrange (1771) [1773], §109.

is a striking analogy here with the work of Cardano, who in 1545 had initiated a similarly profound change in perception, away from the collecting of recipes towards an understanding of transformations of equations. Both Cardano and Lagrange gathered the best knowledge of their time on equation-solving and both were thereby led to insights that pushed the discussion to a more abstract and more challenging level.

Lagrange had not, as he had hoped, put to rest the problem of equation-solving but had instead opened up an entirely new line of research: the investigation of functions of the roots and of the number of values such functions could take as the roots were permuted. In the hands of Cauchy and Galois in the early nineteenth century such research was to lead directly to the foundations of modern group theory, and to a radical transformation of what algebra itself was perceived to be.

Chapter 11
The outsiders: Waring and Vandermonde

Almost all of the work described in the last six chapters was done by just three men: Euler, Bezout, and Lagrange, based in St Petersburg, Paris, and Berlin. None of them ever met each other personally but they communicated through the journals of their respective Academies. In this chapter we look at two other mathematicians who investigated similar themes but whose work fell outside the mainstream of mid-eighteenth-century mathematical activity: Edward Waring in Cambridge and his exact contemporary Alexandre-Théophile Vandermonde in Paris. Both were particularly active around 1770, just as Lagrange too was turning his attention to equations. An investigation of Waring's and Vandermonde's interactions, or rather the lack of them, with the mathematicians named above reveals parallel but independent development of similar mathematical ideas by people unconnected to each other. This phenomenon is by no means unknown in mathematics but this particular example has not, to my knowledge, previously been highlighted.

The end of this chapter offers a summary of the key insights into equation-solving up to 1771.

Waring's *Meditationes algebraicae*, 1770

Edward Waring was born into a farming family in Shrewsbury and entered Magdalen College, Cambridge, in 1753. He began working on a treatise known as *Miscellanea analytica* at least as early as 1757, when part of it was submitted to the Royal Society. In December 1759, the death of John Colson opened up a vacancy for the Lucasian chair, and Waring, who had graduated as BA but not yet MA, became a candidate. In support of his application he printed and circulated a few copies of the first chapter of the *Miscellanea*. It was severely criticized by William Powell, a tutor at St John's College (who favoured a different candidate), but was ably supported by John Wilson, then an undergraduate at Peterhouse, with the result that Waring was appointed to the Lucasian Chair in January 1760. The full text of the *Miscellanea* was published at Cambridge in 1762. The first half, on the theory of equations (65 pages), was subsequently greatly expanded by Waring and was republished in what he called a 'second edition' as the *Meditationes algebraicae* (219 pages) in 1770. The third and final edition appeared in 1782.

Waring's *Meditationes* are aptly named. The book has all the qualities of a fertile but wandering mind: ideas arise, intermingle, and coalesce apparently at random, appearing brilliant for a time but then subsiding into obscurity or lengthy algebraic calculations. The usual structures of good mathematical writing are entirely missing: there is no sense here of building from basic principles or easy examples to more general theorems. Instead, problems, lemmas, corollaries, and examples tumble over one another without apparent order or reason so that the reader is left without any sense of either starting point or direction. As the anonymous editor of *The Georgian era* later wrote under the entry for Edward Waring:[1]

> The reader [...] is stopped at every instant, first to make out the author's meaning, and then to fill up some chasm in the demonstration. He must invent anew every invention; for after the enunciation of the theorem or a problem, and the mention of a few leading steps, little farther assistance is afforded.

All the same difficulties had plagued the *Miscellanea* of 1762. This first edition had boasted a list of some 320 subscribers, most of them from Cambridge colleges or from Waring's native Shropshire, but few can have been able to follow Waring beyond the first few pages.

The *Miscellanea* consists of five chapters. Chapter I poses the problem of finding the equation whose roots are some algebraic function of the roots of a given equation (Problem I); in order to answer this Waring set up a number of formulae for sums of powers and sums of other rational functions of the roots. Chapter II takes up the idea first explored by Newton of using such formulae to set upper and lower bounds for the roots; it also discusses the inadequacy of the known rules for 'impossible roots'. Chapter III investigates equations whose roots are in some particular relation to each other (in arithmetic progression or geometric progression, for example) (Problem II); it also explores another problem: given a polynomial equation in x and y, find x and y as rational functions of some third variable z (this is called 'Problem IV' but is actually the third of Waring's 'Problems'). Chapter IV examines the degree of an equation by which an equation of degree n can be reduced to an equation of degree m where $m < n$ (Problem V). The phrasing of the problem does not make clear what Waring had in mind, but he was thinking of what he called a 'reducing equation' (*aequatio reducens*) and what continental writers called a 'resolvent' (*aequatio resoluens).* He argued that in general the degree of such an equation was going to be higher than that of the original equation and therefore that such methods were useless in the search for general solutions. Finally, Chapter V looks at the problem of reducing two equations to one by elimination (Problem IX); the constitution of the coefficients in equations with more than one unknown (Problem X); transforming an equation in two unknowns to another with roots in a given relation to the roots of the original (Problem XI); and so on.

It is clear even from this brief summary that Waring was concerned with many of the same problems that occupied Euler from time to time during the 1740s and 1750s, and both Bezout and Euler in the early 1760s. It is therefore pertinent to ask what interactions, if any, existed between them. When Jérôme Lalande in his 'Notice sur la vie de Condorcet' (1796) asserted that in 1764 there had been no first-rate analyst in England,[2] Waring pointed out that in pure mathematics he himself had contributed

[1] *The Georgian era*, 1832–34, III, 200.
[2] Lalande 1796.

'somewhere between three and four hundred new propositions of one kind or another' and that both d'Alembert and Lagrange had mentioned the *Meditationes* as 'a book full of interesting and excellent discoveries in algebra'. At the same time he asserted with some pride that d'Alembert, Euler, and Lagrange had published discoveries that they might possibly have seen in his *Miscellanea*:[3]

> I must congratulate myself that D'Alembert, Euler, and Le Grange, three of the greatest men in pure mathematics of this or any other age, have since published and demonstrated some of the propositions contained in my Meditationes Algebraicae or Miscellanea Analytica, the only book of mine they could have seen at the time.

It is true that Euler and Lagrange trod some of the same ground as Waring, but his suggestion that they saw their results first in his writings does not stand up to scrutiny.

Waring had indeed sent his *Miscellanea* to Euler early in 1763, but whether Euler read it we do not know. Waring later pointed out that in the *Miscellanea* he had suggested that solutions to n^{th}-degree equations might take the form $x = a\sqrt[n]{p} + b\sqrt[n]{p^2} + c\sqrt[n]{p^3} + \cdots + D\sqrt[n]{p^{n-1}}$, and that both Euler and Bezout had afterwards published the same suggestion (both in 1764).[4] Euler, however, had first moved towards this idea some thirty years earlier in 1733 and had then written about it again in 1753 leading eventually to the published version in 1764; Bezout, by his own admission, took up the theme from Euler and had begun to work on it in 1762 before he could have seen Waring's *Miscellanea*. There is therefore nothing to suggest that the *Miscellanea* influenced either Euler or Bezout in their researches during the early 1760s.

The *Meditationes* of 1770 was circulated more widely: Waring sent it in May of that year to d'Alembert, Bezout, Euler, and Lagrange, but complained in 1782 that none of them had acknowledged it.[5] Lagrange, however, referred to it twice with some admiration in the final section of his 'Réflexions' written in 1771 and 1772. As he had not mentioned it in the historical introduction at the beginning of his paper we may suppose that he read it only as he was completing the work in late 1771 or very early in 1772. Lagrange commented in particular on Waring, alongside Cramer, for his work on rational functions of the coefficients of an equation, and described the *Meditationes* as 'a work full of excellent research on equations'.[6] Towards the end of his paper Lagrange also discussed equations that could be reduced in degree because of some special relationship between the roots. Lagrange believed that Hudde had been the first to discuss such cases but remarked that many later geometers had also dealt with them, above all Waring in 'the excellent work cited above'.[7] Lagrange, as

[3]Cited in *The Georgian era*, 1832–34, III, 199. See also Waring 1799.

[4]Waring 1782, xxi.

[5]Waring 1782, xxi.

[6]*Voyez là-dessus, outre l'Ouvrage de M. Cramer que nous avons déjà cité, encore celui de M. Waring, qui a pour titre Meditationes algebräicae, Ouvrage rempli d'excellentes recherches sur les équations*. Lagrange 1773, §96.

[7]*D'autres Géometres* [...] *ont perfectionné et étendu plus loin les regles et les méthodes de M. Hudde; (voyez surtout l'excellent Ouvrage de M. Waring cité ci-dessus).* Lagrange (1771) [1773], §110.

always, was generous in acknowledging the work of his predecessors and his contemporaries, but though he may have recognized the quality and correctness of Waring's results he can have learned little that he did not know already, for he had already covered much of the same ground himself, and much more systematically than Waring had done. Vandermonde in 1774 also referred to Waring's *Meditationes*, but as a book he had seen only after he had discovered certain results for himself (see below). Just as for the *Miscellanea* earlier, therefore, there is nothing to suggest that Waring's *Meditationes* had any significant influence on the research of continental mathematicians.

We must also ask the converse question: how much did Waring in Cambridge in the 1760s know of the research being done in Paris or Berlin? Waring himself provides the answers because he began all his books with historical prefaces outlining results achieved up to then. In the *Miscellanea* of 1762 he referred to many of his seventeenth-century predecessors: Viète, Descartes, Harriot, Oughtred, van Schooten, Wassenaer, Hudde, and Bartholin, but when it came to the eighteenth century there were just two: Cramer and Newton. The names enable us to reconstruct a list of books that Waring probably read as a young man: Oughtred's *Clavis* and Harriot's *Praxis* (both published in 1631); van Schooten's editions of Viète's *Opera mathematica* (1646) and Descartes' *Geometria* (1659–61) with its additional papers by van Schooten, Hudde, and Bartholin; Cramer's *L'analyse des lignes courbes* (1750); and Newton's *Arithmetica universalis* (1707).

It is clear from the ideas that Waring explored in the *Miscellanea* that he was particularly indebted to the *Arithmetica universalis*, perhaps not surprisingly given his residence in Cambridge. Chapter I of the *Miscellanea*, for instance, opens with formulae for sums of powers of roots of an equation, derived and extended from those given by Newton in the *Arithmetica universalis*.[8] Chapter II is concerned with approximations to the roots, based on Newton's similar approximations using sums of powers at the end of the *Arithmetica universalis*;[9] and with Newton's rule for the number of 'impossible' roots.[10] In Chapter IV, Waring asserted, among other things, that the degree of a 'reducing equation' (*aequatio reducens*) may be of higher degree than that of the original equation, with a specific reference to the *Arithmetica universalis*.[11] Chapter V, the last, begins with yet another problem raised by Newton, that of reducing two equations to one by elimination.[12] Thus there is ample evidence to suggest that Waring as a student found in the *Arithmetica universalis* alone the seeds of what became his own wild and overgrown garden.

---

[8]Newton 1707, 251–252; 1720, 205–206.

[9]Newton 1707, 252–257; 1720, 206–210.

[10]Newton 1707, 242–245; 1720, 197–200.

[11]Waring 1762, 34. The reference is '*Arith. Univ*. p. 237' but this is clearly wrong since p. 237 in the 1707 edition of the *Arithmetica universalis* (the edition that Waring used) does not contain any relevant results. What Waring seems to have in mind is Newton 1707, 272–276 (perhaps 237 was a misprint for 273?) where Newton showed that a cubic can be resolved by means of a quadratic and a quartic by means of a cubic. Waring's claim appeared again in Waring 1770, 89, but this time the reference to Newton was omitted.

[12]Newton 1707, 69–76; 1720, 60–67.

By the time his early writings became absorbed into the much longer *Meditationes* of 1770, Waring's horizons had expanded. His preface now offered a much longer and more detailed historical introduction, based partly on Wallis's *A treatise of algebra* (1685) and Montucla's two-volume *Histoire des mathématiques* (1758), but it also showed greater familiarity with the writings of his continental contemporaries. By now, for instance, he knew of Euler's 'Recherches sur les racines imaginaires des equations' (1749) [1751] in which Euler had first suggested constructing an equation in a function of the roots, but Waring said that he had not seen it until after the *Miscellanea* was published.[13] He also knew of some of Euler's work on elimination,[14] presumably his 'Nouvelle méthode d'éliminer les quantités inconnues des equations' (1764) [1766]. Both of these papers had appeared in the *Mémoires* of the Berlin Academy. Further, Waring knew of Bezout's exploration of roots as sums of radicals (1762) [1764], which had appeared in the *Mémoires* of the Paris Academy, but apparently not yet of Euler's paper on the same subject (1762–63) [1764] which had appeared in the *Novi commentarii*, and which may not have been available to him in Cambridge.[15] He referred to this last paper much later, in the Preface to his 'third edition' in 1782, when he pointed out that he had sent his *Miscellanea* to Euler in 1763.[16]

Thus it appears that Waring became only slowly aware of the work of Euler and Bezout, and only after he himself had discovered many similar results for himself. His findings and theirs thus seem to have been genuinely independent. The same can be said of Waring and Lagrange, for Lagrange appears to have read some of Waring's results only in late 1771 *after* he had completed his own long study of equations. Waring was proud of his own achievements but never seriously complained that others had pre-empted or plagiarized him, only that they had published similar results.

In short, during the 1760s, Waring, Euler, Bezout, and later Lagrange worked on very similar themes, but only the last three were really aware of each other. Waring in England was very much on his own.

Vandermonde's paper of 1770

The case of Alexandre-Théophile Vandermonde in relation to Lagrange and earlier algebraists is less complicated than that of Waring. Until 1770 Vandermonde followed a career as a violinist. What made him then turn to mathematics is not clear. However, his first paper, 'Mémoire sur la resolution des équations', shows that he was familiar with Euler's paper on roots as sums of radicals (1762–63) [1764], and Bezout's on the degree of the resolvent (1765) [1768], suggesting that his understanding of mathematics was already quite sophisticated. Vandermonde's paper on equations was presented to the Paris Academy in November 1770, but publication had to wait until he became a member in May the following year. The volume of *Mémoires* for 1771 was not printed until 1774, by which time Waring's *Meditationes* and Lagrange's *Réflexions* had also

[13]Waring 1770, iv–v.
[14]Waring 1770, v.
[15]Waring 1770, v.
[16]Waring 1782, xxi.

appeared. In a footnote Vandermonde noted the existence of both and remarked that he could only be flattered that these authors had discovered some results similar to his own.

Vandermonde's 'Mémoire' did indeed go over some of the same ground that Lagrange and, more particularly, Waring had covered before him. He began by noting that a root of the equation

$$x^2 - (a+b)x + ab = 0$$

must be a function of $(a+b)$ and of ab, both of which remain the same if a and b are interchanged. Thus we must seek a function of $(a+b)$ and ab that takes two values, so that, as Vandermonde wrote it

$$a = \textit{fonction}[(a+b), ab]$$

and

$$b = \textit{fonction}[(a+b), ab].$$

Such a function might be, for instance,

$$\tfrac{1}{2}[(a+b) + \surd((a+b)^2 - 4ab)]$$

or

$$\tfrac{1}{2}[(a+b) + \tfrac{\surd(2\surd-1)+\surd(-2\surd-1)}{2}\surd((a+b)^2 - 4ab)]$$

or

$$\frac{2ab}{(a+b) + \surd((a+b)^2 - 4ab)}.$$

The first of these (which is the usual formula for a quadratic equation) is the simplest, and was therefore, in Vandermonde's view, the most useful. In each of the above expressions the square root of $(a+b)^2 - 4ab$ introduces an ambiguity which makes the value of the function either a or b.

Turning to a cubic equation with roots a, b, c, Vandermonde argued by analogy that an appropriate function might be

$$\tfrac{1}{3}[a+b+c+\sqrt[3]{(a+r'b+r''c)^3} + \sqrt[3]{(a+r''b+r'c)^3}] \tag{1}$$

where $1, r', r''$ are the cube roots of 1. It is easy to see that

$$\begin{aligned}\tfrac{1}{3}[a+b+c+(a+r'b+r''c)+(a+r''b+r'c)] &= a,\\ \tfrac{1}{3}[a+b+c+r''(a+r'b+r''c)+r'(a+r''b+r'c)] &= b,\\ \tfrac{1}{3}[a+b+c+r'(a+r'b+r''c)+r''(a+r''b+r'c)] &= c,\end{aligned}$$

as required. It therefore remained to prove that $(a+r'b+r''c)^3$ and $(a+r''b+r'c)^3$ are functions of $a+b+c$ and $ab+ac+bc$ and abc only.[17] Now

$$\begin{aligned}(a+r'b+r''c)^3 = a^3+b^3+c^3 \\ -\tfrac{3}{2}(a^2b+a^2c+b^2a+b^2c+c^2a+c^2b)+6abc \quad (2)\\ +\tfrac{3}{2}(a^2b+b^2c+c^2a-a^2c-b^2a-c^2b)\sqrt{-3}.\end{aligned}$$

This is unchanged by permutations of a, b, c, except for the final term which can take just two values, differing only in sign. This can be seen by inspection but Vandermonde also observed that the final term is a multiple of

$$(a^2b+b^2c+c^2a-a^2c-b^2a-c^2b) = -(a-b)(b-c)(c-a), \qquad (3)$$

which makes it clear that any transposition of the letters gives rise only to a change of sign.[18] Vandermonde calculated the square of (3) as

$$\begin{aligned}&a^4b^2+a^4c^2+b^4a^2+b^4c^2+c^4a^2+c^4b^2\\ &-2(a^4bc+b^4ac+c^4ab)-2(a^3b^3+a^3c^3+b^3c^3)\\ &+2(a^3b^2c+a^3c^2b+b^3a^2c+b^3c^2a+c^3a^2b+c^3b^2a)-6a^2b^2c^2.\end{aligned}$$

It is not hard to see that this is indeed invariant under permutations of a, b, c. Vandermonde claimed an apparently stronger condition, that it was in fact, expressible in terms of $a+b+c$ and $ab+ac+bc$ and abc, as he would shortly demonstrate. These observations led Vandermonde to the following conclusions as to how to solve an equation of any degree:[19]

1. Find a function of the roots, of which one may say, in a certain sense, that it is equal to each of the required roots.

2. Put this function into a form that remains mostly unchanged when the roots are exchanged between themselves.

3. Write the values in terms of the sum of the roots, the sum of their products in pairs, and so on.

---

[17] Vandermonde's $(a+r'b+r''c)^3$ and $(a+r''b+r'c)^3$ here are the same as Lagrange's $(x_1+\alpha x_2+\alpha^2x_3)^3 = t_1^3 = \theta_1$ and $(x_1+\alpha^2x_2+\alpha x_3)^3 = t_4^3 = \theta_2$, see page 178.

[18] The expression $(a-b)(b-c)(c-a)$ on the right is the value of what later became known as the 'Vandermonde determinant', $\begin{vmatrix}1&1&1\\a&b&c\\a^2&b^2&c^2\end{vmatrix}$, but no determinant in this form appeared in Vandermonde's paper. Vandermonde himself did much to develop the theory of determinants but not until later, in Vandermonde (1772b) [1776]. The name 'determinant' was not given to such arrays until 1815, by Cauchy.

[19] 1.º *Trouver une fonction des racines, de laquelle on puisse dire, dans un certain sens, qu'elle égale telle de ces racines que l'on voudra.*

2.º *Mettre cette fonction sous une forme telle qu'il soit de plus indifférent d'y échanger les racines entr'elles.*

3.º *Y substituer les valeurs en somme de ces racines, somme de leurs produits deux-à-deux, &c.* Vandermonde (1771a) [1774], §IV, 370.

370 MÉMOIRES DE L'ACADÉMIE ROYALE

Voilà donc la ſeconde condition de la réſolution générale du troiſième degré ſatisfaite. Quant à la troiſième & dernière qui eſt de ſubſtituer à toutes ces quantités, dans leſquelles il eſt indifférent d'échanger les lettres entr'elles, leurs valeurs en. . . . $(a+b+c)$, $(ab+ac+bc)$ & abc, on verra dans l'inſtant qu'elle eſt toujours très-facile à remplir.

IV. On voit dès-à-préſent que pour un degré quelconque, la condition eſſentielle de la réſolution générale étant de trouver une fonction de la ſomme des racines, de la ſomme de leurs produits deux-à-deux, de la ſomme de leurs produits trois-à-trois, &c. qui ſoit indifféremment l'une quelconque de ces racines, cette recherche peut ſe partager en trois chefs:

1.° *Trouver une fonction des racines, de laquelle on puiſſe dire, dans un certain ſens, qu'elle égale telle de ces racines que l'on voudra.*

2.° *Mettre cette fonction ſous une forme telle qu'il ſoit de plus indifférent d'y échanger les racines entr'elles.*

3.° *Y ſubſtituer les valeurs en ſomme de ces racines, ſomme de leurs produits deux-à-deux, &c.*

Traitons d'abord ce troiſième objet qui eſt le plus ſimple.

V. J'écrirai dans la ſuite

(A) pour $a+b+c+$&c. (A^2) pour $a^2+b^2+c^2+$&c.

(AB) pour $ab+ac+$&c. (A^2B) pour $a^2b+b^2a+c^2a+$&c.

$+bc+$&c. $+a^2c+b^2c+c^2b+$&c.

$+$&c. $+$&c. $+$&c. $+$&c.

&c.

En général par $(A^\alpha B^\beta C^\gamma \ldots)$ j'indiquerai la ſomme de tous les termes différens qui réſulteroient de celui-là, au moyen de toutes les ſubſtitutions poſſibles des petites lettres $a, b, c, d, e,$ &c. dans un ordre quelconque, aux grandes, $A, B, C,$ &c. & j'appellerai ces expreſſions abrégées, *Types de combinaiſon* ou ſimplement *Types.*

Comme le cas où quelques-uns des expoſans $\alpha, \beta, \gamma,$ &c.

Three steps for solving equations, from Vandermonde (1771).

For a cubic equation, the first of these problems was addressed in equation (1), where it can be seen that appropriate choices of cube root lead to each root of the equation in turn. The second point is illustrated by (2), which for the most part remains unchanged under permutations of a, b, c, although the final term can take two values (differing only in sign). The third point is demonstrated by (3) whose square, according to Vandermonde, can be calculated in terms of $a + b + c$ and $ab + ac + bc$ and abc only, that is, in terms of the coefficients of the original equation.

In outlining the extension of his theory to equations of any degree Vandermonde decided to begin with the third of these problems, which he regarded as the simplest. He therefore introduced the notation

$$\begin{array}{lcl} \{1\} = (A) &=& a + b + c + \cdots, \\ \{2\} = (A^2) &=& a^2 + b^2 + c^2 + \cdots, \\ \{1^2\} = (AB) &=& ab + ac + bc + \cdots, \\ \{21\} = (A^2B) &=& a^2b + a^2c + \cdots + b^2c + \cdots, \\ \ldots \end{array}$$

and in general

$$\{\alpha\beta\gamma\ldots\} = (A^\alpha B^\beta C^\gamma \ldots)$$

for the sum of terms of the form $a^\alpha b^\beta c^\gamma \ldots$. These various sums he called 'types of combination' (*types de combinaison*) or simply 'types' (*types*). The next few pages of his paper are taken up with establishing relationships between types, for example,

$$\{5\} = \frac{5.1.2.3.4}{1.2.3.4.5}\{1\}^5 + \frac{5.1.2.3}{1.2.3}\{1\}^3\{1^2\} + \frac{5.1.2}{1.2}\{1\}\{1^2\}^2 + \frac{5.1.2}{1.2}\{1\}^2\{1^3\}$$
$$-\frac{5.1}{1}\{1^2\}\{1^3\} - \frac{5.1}{1}\{1\}\{1^4\} + \frac{5}{1}\{1^5\},$$

or, in his alternative notation,

$$(A^5) = (A)^5 - 5(A)^3(AB) + 5(A)(AB)^2 + 5(A)^2(ABC)$$
$$- 5(AB)(ABC) - 5(A)(ABCD) + 5(ABCDE).$$

The published version of his paper contains a large fold-out table of such relationships.[20]

Vandermonde turned next to the first of his three steps and by analogy with his previous results proposed the general function

$$\frac{1}{n}\Bigg[a + b + c + \cdots + \sqrt[n]{(a + r_1b + r_2c + \cdots)^n} + \sqrt[n]{(a + r_1^2b + r_2^2c + \cdots)^n}$$
$$+ \cdots + \sqrt[n]{(a + r_1^{n-1}b + r_2^{n-1}c + \cdots)^n}\Bigg],$$

[20] Euler had derived similar relationships in Euler (1770) [1771] (see Chapter 6) but the near simultaneous composition of their papers in 1770 meant that neither Euler nor Vandermonde could have known of each other's results.

where $1, r_1, r_2, \ldots, r_{n-1}$ are n^{th} roots of 1. Vandermonde demonstrated at length how this function can be made equal to each of the roots $a, b, c, \ldots$ when $n = 4, 5, 6$, or 7. He also noted that in cases where n is non-prime some modifications of the basic function are possible, a hypothesis he tested further for $n = 8$ and $n = 9$, and he claimed that such simplifications arise from additional symmetries between the roots (§XVIII). He observed, however, that there are essentially only two different forms of the function when $n = 4$ and only three when $n = 6$. Since his aim was to find general rules rather than simplifications in particular cases, he declined to discuss this aspect any further.

Now, Vandermonde claimed, it remained only to work on the second of the three steps, namely to transform this general function into a function of types. Thus, for example, returning to the case $n = 3$, the function in (1) can be written as

$$\frac{1}{3}\Big[(A) + \sqrt[3]{(A^3) - \tfrac{3}{2}(A^2B) + 6(ABC) + \tfrac{3}{2}(a-b)(a-c)(b-c)\sqrt{-3}}$$
$$+ \sqrt[3]{(A^3) - \tfrac{3}{2}(A^2B) + 6(ABC) - \tfrac{3}{2}(a-b)(a-c)(b-c)\sqrt{-3}}\Big].$$

Although $(a-b)(a-c)(b-c)$ takes two values under permutations of a, b, c, they differ only in sign. In fact, as we saw above following equation (3), $(a-b)(a-c)(b-c)$ is the square root of a function of types, namely

$$(A^4B^2) - 2(A^4BC) - 2(A^3B^3) + 2(A^3B^2C) - 6(A^2B^2C^2).$$

For $n = 3$ all the types could be calculated with the help of his fold-out tables, leading to the usual well known solutions for a cubic. Vandermonde performed a similar calculation for $n = 4$, again arriving at the usual solutions (§XXI). For $n = 5$ his calculations became exceedingly complex, and he observed that the eventual equation (which, he noted, other authors called either *résolvante* or *réduite*) would be of degree 24. Its coefficients, he claimed, would be rational functions of the coefficients of the original equation, calculated from what he called partial types (*types partiels*) (§XXIX, §XXX). His combinatorial arguments here were very similar to those Cramer had proposed in 1750 (see page 139). Finally, for $n = 6$, calculations of similar length led him to the conclusion that (as Hudde had seen more than a century earlier, see page 54) the resolvent would be of degree 10 if the equation is to be expressed as a product of two cubics, or of degree 15 if it is to be expressed as a product of three quadratics (§XXXII).

Vandermonde concluded his paper by returning to the three crucial steps he had outlined near the beginning. The first and the third, he said, were always possible; as for the second he had shown a way forward and he ended optimistically, suggesting that the calculations contained no more difficulty than the inevitable length. Thus Vandermonde joined all the other writers (Tschirnhaus, Euler, Bezout, Lagrange) whose methods and analysis worked beautifully for cubics and quartics but collapsed in a tangle of calculations as soon as they were applied to quintics.

Nevertheless, Vandermonde's achievements were remarkable. Lacking either Lagrange's historical knowledge or mathematical experience he had arrived in a single paper at many similar conclusions. Not the least of these was the insight that solving equations depended upon finding a suitable function of the roots. For Lagrange this had emerged only after lengthy exploration and comparison of all known methods, but Vandermonde seems to have been able to intuit it almost immediately, so that for him it was not an end result but his starting point.

It is clear that Vandermonde was a gifted mathematician, but his flame burned very briefly. His second mathematical paper, also published in the *Mémoires* of the Paris Academy for 1771, took up ideas of Leibniz on *geometria situs*, a geometry of position rather than measurement. His third paper, written in 1772, extended the definition of the function

$$[p]^n = p(p-1)(p-2)\dots(p-n+1)$$

to cases where neither p nor n is an integer. He applied his results to the evaluation of certain integrals and thus rediscovered results on the quadrature of the circle that John Wallis had found (though with much greater labour) in the seventeenth century.[21] Vandermonde's fourth paper, also written in 1772, was on elimination, and in it essentially established a theory of determinants. The first and the fourth papers alone were enough to establish Vandermonde as a clever and innovative mathematician, but there was nothing to follow. Instead he turned to physical experiments and later also to political activity as an ardent supporter of the French Revolution.

Looking back

Here at the end of Part II we may take a moment to look back to some of the key developments in the theory of equations before 1771, before moving forward in Part III to examine the aftermath and influence of the work done by Lagrange and Vandermonde in particular.

With hindsight we can see that many of the themes explored in Part II began to emerge as early as the sixteenth century in the work of Cardano, and in the seventeenth century in the writings of Hudde, Gregory, Tschirnhaus, Leibniz, and Newton. The influence of these writers on mathematicians of the eighteenth century, however, varied greatly. Cardano's name was attached to the rule for cubic equations, but it is unlikely that even in the seventeenth century anyone turned to the *Ars magna* itself as a source of inspiration. Hudde's work, on the other hand was well known because of its publication alongside Descartes' *Geometria* in 1659; it was closely read and admired by Gregory, Tschirnhaus, Leibniz, and Newton, and later also by Lagrange. The findings of Gregory, Tschirnhaus, and Leibniz in the 1670s, however, remained buried in their private correspondence. Euler could not have known, for example, when he proposed that roots might be expressible as sums of radicals (Chapter 5), that Leibniz

[21] One of Vandermonde's results, for instance, was $\int \frac{\mathrm{d}x}{\sqrt{1-x^2}} = \frac{1}{2}\pi = [\frac{1}{2}]^{\frac{1}{2}}[-\frac{1}{2}]^{-\frac{1}{2}} = \frac{2.2.2.4.4.6....}{1.1.3.3.5.5....}$; another was $2[\frac{1}{2}]^{\frac{1}{2}} = \sqrt{\pi}$. For Wallis's results and methods see Wallis 1656.

had long before suggested something similar. Nor could he know that both Gregory and Leibniz in attempting to solve equations of degree 5 had run into equations of much higher degree (Chapter 8); he could have seen the same thing in Hudde, but here as elsewhere Euler seems not to have been particularly well read in seventeenth-century mathematics.

Had it not been for William Whiston's publication of Newton's algebra notes as the *Arithmetica universalis*, Newton's thoughts on equations might also have remained unread. Newton's assertions on the number of imaginary roots of an equation (Chapter 4), on symmetric functions of the roots (Chapter 6), and on elimination (Chapter 7), were all further developed by others. The last two were directly taken up by Euler, though he does not seem to have read the *Arithmetica universalis* until perhaps as late as 1746 by which time he had also begun to work on elimination independently. Further important ideas stemmed from other writings by Newton: his method for the numerical solution of equations (Chapter 9), examined by Lagrange; and, perhaps most important of all, his infinite series for sines and cosines of multiple angles, which were the key to de Moivre's paper of 1707 (Chapter 5). Thus Newton's legacy can be detected at many points in the story but it cannot be claimed that it was pervasive. The motivation that drove de Moivre, Euler, and Bezout to seek out algebraic solutions of equations of higher degree was not Newton's, who never pursued the matter in depth. If he had, the theory might have evolved very much more rapidly than it did.

De Moivre's 'Aequationum quarundam potestatis [...] resolutio analytica' (1707), an extension of Cardano's solution for cubics to certain higher degree equations, was a poorly written paper that offered the reader little in the way of explanation, but can now be seen to have marked a transition in the theory of equations, from a series of scattered results to a more systematic attempt to examine which equations could or could not be solved by a given method. It was the paper that inspired Euler in 1733 to conjecture that the roots of *any* polynomial equation might be expressible as a sum of radicals (Chapter 5). This was an idea that took many years to bear fruit, but both Euler and Bezout eventually pursued it further and their respective publications appeared simultaneously in 1764. Their understanding of roots as sums of radicals proved to be crucial, leading to the most promising transformations of equations since Tschirnhaus, yet the motivation of Euler and Bezout remained much the same as de Moivre's had been in 1707: given the difficulties of solving higher degree equations in general, to find particular classes of equations that *could* be solved algebraically.

In the meantime, other insights were also beginning to come into play. One of the most important was the idea of constructing a secondary equation whose roots were functions of the roots of the equation one wished to solve (Chapter 6). Again, this was suggested first by Euler, who was interested initially in sums of pairs of the roots, and later in the squares of the differences. For Euler, as for Lagrange later (Chapter 9), this was related to efforts to identify the existence of imaginary roots. Eventually, however, such functions came to be regarded as vitally important for other reasons, especially functions that took only a small number of values as the roots of the original equation were permuted (Chapters 9, 10).

The secondary equations satisfied by such functions became known as 'reduced' or 'resolvent' equations. What every equation-solver hoped for was a resolvent of lower degree than the original, satisfied by a function from which the roots of the original equation could be recovered. For cubics and quartics, such resolvent equations were found repeatedly and by a variety of methods; Lagrange was able to show that all of them arose from a restricted number of possible functions (Chapter 10). For quintics and equations of higher degree, however, suitable resolvents remained stubbornly elusive, and where they could be found at all they were invariably of higher degree than the original equation (Chapters 8, 10).

All of this material and more was eventually brought together in Lagrange's *Réflexions* of 1771, the culmination of progress on equations in the eighteenth century. In the final chapter of this book we will see how Lagrange's work led in his lifetime and beyond to changes in the nature of algebra itself.

Part III

After Lagrange

Chapter 12
Dissemination and new directions

This final chapter looks at the dissemination of the results derived by Euler and Bezout in the 1750s and 1760s and by Lagrange, Waring, and Vandermonde in the early 1770s, not only to mathematicians in academies and universities who might pursue further research but also to a more general readership. Both Euler and Bezout were active teachers of mathematics, and so it is perhaps not surprising that the earliest elementary expositions of their new ideas were to be found in their own textbooks. Euler's textbook treatment of equations did not appear until many years after his initial findings, but for Bezout, textbook and Academy publication were almost simultaneous.

Lagrange's work, on the other hand, was much less amenable to elementary treatment and was taken up only by professional mathematicians. The first of these was his fellow Italian Ruffini, followed later by Cauchy, Abel, and Galois. In their hands, Lagrange's results of the 1770s led in two distinct but related directions: first, towards a proof of the general insolvability of quintic and higher degree equations; second towards the founding of a completely new branch of algebra, the theory of permutation groups.

Euler's *Elements of algebra* and Bezout's *Cours de mathématiques*

Euler's *Elements of algebra* became one of the most widely used algebra texts of the late eighteenth century. First published in Russian in 1768–69, it was translated into German in 1770, into French in 1774, and into English in 1797. To bypass the various changes of title in these several languages we will keep here to the English title.[1] Only a little of Euler's original thinking on equations found its way into this book, which was aimed very much at beginners. Equations and their solutions were not treated in detail until the final section of Book I where Euler, in typically sound pedagogic fashion, worked gradually upwards through the standard methods of solving equations of degree 1, 2, 3, and 4. Only after that did he turn to what he called 'a new method of resolving equations of the fourth degree', where he introduced the idea that the root of such an equation might be of the form $\sqrt{p}+\sqrt{q}+\sqrt{r}$ where p, q, r are the three roots of a cubic equation. 'New' has to be regarded as a relative term since Euler had first made this suggestion well over thirty years earlier. His exposition ended with three well chosen examples for the student to work for himself.

Bezout's *Cours de mathématiques, à l'usage des Gardes du Pavillon et de la Marine*, first published in 1764–66, went much further. Intended as a teaching text for the young naval students in his charge, it was written in six parts: (i) arithmetic, (ii) geometry and trigonometry, (iii) algebra, (iv) principles of calculus and mechanics, (v) applications of

[1] For further details of the various translations see the bibliography.

those principles, (vi) navigation. Initially these were printed separately and complete books were sometimes made up by binding together parts from different print runs. A parallel *Cours* for the artillery was also published from 1770 onwards, containing the same material apart from the section on navigation.[2] The section on algebra is detailed and comprehensive, covering systems of linear equations; quadratic equations; the binomial theorem for an integer power; elimination of one of two unknowns from equations of degree greater than 1; composition of polynomials as products of factors; transformation of equations and in particular removal of the second term; the solution of equations of degree 3 or 4 by substituting a variable y with the property $y^m - 1 = 0$; finding common divisors; solution by approximation; finding equal or imaginary roots. All this, one might think, could be more than a fledgling naval or artillery officer could easily handle, so Bezout helpfully separated out what he regarded as the more difficult themes and set them in smaller print. The small print starts to creep into the two sections on the binomial theorem and elimination, and is used for everything from the composition of polynomials onwards.

In his preface to the first edition of the algebra section, published in 1766, Bezout highlighted two topics in particular. The first was the problem of elimination which, he said, one would find explored at greater length in his article in the *Mémoires* of the Paris Academy.[3] He noted, however, that there was still much to be done. Eventually, a footnote added to a later printing, in 1781, claimed that this was no longer the case after the publication of his *Théorie générale des équations algébriques* in 1779.[4]

The second subject to which Bezout drew attention in 1766 was the search for a general method of solving equations of any degree. On this, he said, he would write nothing about methods that had been tried up to then except that none of them worked beyond degree 4. He had not intended to publish his own work until it was perfected, but Euler had recently come out (in 1764) with similar results in *Novi commentarii* 9. Bezout was therefore publishing in his *Cours* what he himself had found up to 1761. For more detail the reader was referred to the *Mémoires* of the Paris Academy.[5] Thus Bezout's early research appeared not only in the *Mémoires* but simultaneously in a widely read elementary textbook. Indeed, his method of reducing an equation to the form $y^m - 1 = 0$ was published in his *Cours mathématiques* even before it appeared in the *Mémoires*.

Comparing Bezout's textbook exposition with Euler's, we see that Euler treated ideas that were earlier than Bezout's and that he explained them at a much more

[2]The *Cours de mathématiques* was reprinted many times. A second edition, using post-Revolution units, was used by the École Polytechnique from 1798 onwards, followed by a third, augmented, edition after 1809. The parallel *Cours* for the artillery was published in four parts: (i) arithmetic, geometry, trigonometry, (ii) algebra, (iii) principles of calculus and mechanics, (iv) applications. This too was reprinted several times until a single edition for the marines and the artillery replaced previous versions in 1822. For further details see the bibliography.

[3]Presumably Bezout (1764) [1767].

[4]Bezout 1781–84, III, vii.

[5]Presumably Bezout (1762) [1764], already published, and Bezout (1765) [1768], as yet forthcoming. Bezout's method of using the equation $y^m - 1 = 0$, which is described in his *Cours*, was extensively treated in the latter.

elementary level. The differences stem from their respective motivations: Euler was concerned only to give his students a selection of workable methods whereas Bezout was laying claim to new ideas and asserting independent discovery.

The ideas that were to be most influential in terms of later research, however, were not to be those of Euler or Bezout, but those that Lagrange explored in the two final sections of his 'Réflexions', concerning functions of the roots and the effects of permuting the variables they contained. The work that stemmed from such investigations goes well beyond the scope of this book but is summarized briefly here to give some idea of how the theory of equations of the eighteenth century was received and transformed in the nineteenth.

Ruffini and Abel and the insolubility of the quintic

The first mathematician to explore Lagrange's ideas in depth was Paolo Ruffini, from Modena in northern Italy. Shortly before his twenty-third birthday, in 1788, Ruffini graduated from the University of Modena in philosophy, medicine, and mathematics. Three years later he was licensed to practise medicine but also continued to teach mathematics. From 1798 to 1814, during Napoleon's occupation of northern Italy, he was excluded from public office for refusing to swear allegiance to the French Republic. During these years he survived by practising medicine but also did his most important work in mathematics. In 1799 he published his *Teoria generale delle equazioni* in which he claimed to have proved that equations of degree five cannot be solved algebraically.

Ruffini's contemporaries were not immediately convinced by his proof, not least because his exposition was so difficult to follow. Indeed the question of whether Ruffini did or did not prove the insolvability of quintics has continued to perplex modern historians, for reasons we shall examine further below. What was never in doubt, however, was Ruffini's debt to Lagrange, which he himself made clear from the outset: 'The immortal Lagrange with his sublime reflections on equations, inserted into the Acts [*Mémoires*] of the Berlin Academy, has provided the foundation of my demonstrations.'[6] Indeed much of the early part of the *Teoria* is a recapitulation of Lagrange's 'Réflexions', using the same notation. Ruffini went on to show, by explicitly listing and analysing the 120 permutations of five variables, that it is not possible for a function of those variables to take 3 or 4 (or 8) values. He then used these facts to demonstrate that a resolvent equation for a quintic cannot be found.

Here, however, lay the weakness in his argument: Ruffini seems to have assumed that if a resolvent with rational coefficients could not be found then the original equation was unsolvable, whereas in fact all he had proved was that the equation was unsolvable by means of a resolvent with rational coefficients.[7] Ludvig Sylow noted this flaw in

[6] *L'immortale de la Grange con le sublimi sue riflessioni intorno alle equazioni, inserite negli Atti dell'Accademia de Berlino, ha somministrato il fondamento alla mia dimostrazione:* Ruffini 1799, iii.

[7] *Mais bien sûr l'objection majeure subsiste; la démonstration de Ruffini démontre seulement l'impossibilité de résoudre l'équation du cinquième degré par une méthode de transformation-réduction;* [But certainly the main objection remains; Ruffini's proof shows only the impossibility of solving fifth-degree equations by a method of reduction.] Cassinet 1988, 38.

his edition of Abel's papers in 1881; a century later, Raymond Ayoub, Robert A Bryce, and Jean Cassinet, all writing during the 1980s, again identified the same problem but using modern mathematical terminology described it in different ways.[8] The question, therefore, is not whether or not there is a flaw in Ruffini's proof, but whether or not it matters. After all, many first attempts at mathematical proofs contain gaps or errors that have to be sorted out later but which do not necessarily invalidate the entire argument. Ayoub came to the conclusion that, despite its shortcoming, Ruffini's proof essentially succeeded; Bryce and Cassinet remained unconvinced. There the matter must probably rest since, as all readers of Ruffini acknowledge, it is so often impossible to determine precisely what he was trying to say.

At the time, Ruffini did have one supporter. Just a few years before he published his *Teoria*, another Italian professor of mathematics, Pietro Paoli from Pisa, had also come to admire the work of Lagrange and had included some results from the 'Réflexions' in his own *Elementi d'algebra* published in 1794.[9] It is perhaps not surprising that Lagrange's work on equations was of such interest to Italian mathematicians. Not only had Lagrange been born in Turin, but the subject itself had first taken root in northern Italy two centuries earlier. As Paoli wrote in 1804 in the *Supplemento* to the third edition of his *Elementi*:[10]

> It may be observed here that the general resolution of equations, progress in which is owed to the Italian analysts, Scipione Ferri, Tartaglia, Ferrari, Bombelli, has been completed by the work of two Italian geometers: Lagrange and Ruffini.

Some of Ruffini's other compatriots were less convinced of his achievement. Gianfranco Malfatti and Gregorio Fontana, professors of mathematics at Ferrara and Pavia respectively, both raised objections to parts of Ruffini's proof. Meanwhile, Pietro Abbati who, like Ruffini, was based in Modena, offered a completion and clarification of some important details.[11] The outcome of these various discussions was a series of further explanatory papers from Ruffini between 1802 and 1806. The approval he must have desired most, however, was that of Lagrange himself, who of all people might have been expected to understand the proof and confirm it, if it was indeed valid. But from Lagrange there was only silence. Ruffini sent him two copies of the *Teoria*, in 1801 and 1802, but he did not respond. Further, in the introduction to the 1808 edition of his *Traité de la résolution des équations numériques*, Lagrange observed that even if one could find algebraic formulae for solving equations of degree five or higher, they would be of little use for numerical computation.[12] In other words, it seemed that he was not yet convinced that no such formulae could exist.

[8] Abel 1881, II, 293; Ayoub 1980, 265; Bryce 1986, 172–173; Cassinet 1988, 38.

[9] Paoli 1794, I, 119.

[10] *E qui, giova osservare che la risoluzione generale dell'equazioni, i progressi della quale si devono agli Analisti Italiani Scipione Ferri, Tartaglia, Ferrari, Bombelli, he ricevuto il suo compimento per opera di due Italiani Geometri Lagrange e Ruffini.* Paoli, 1804, 127.

[11] For details of the reception of Ruffini's proof in Italy see Cassinet 1987, 38–51.

[12] Lagrange 1808, vii–viii.

Frustrated by the lack of acceptance of his proof, Ruffini wrote in 1808 to Jean Baptiste Joseph Delambre, secretary to the Paris Academy, asking the Academy itself to pass judgement on his work.[13] It was not until April 1810, however, that Lagrange, Legendre, and Lacroix were appointed to examine Ruffini's latest memoir. A year later it was returned to Ruffini without a decision. Lacroix later claimed that he had never even seen it. Meanwhile Lagrange and Legendre had found themselves either unable or unwilling to come to a conclusion.

Every writer on Ruffini admits that his work is obscure, at times impenetrable. Indeed, he has sometimes been compared to Waring thirty years earlier: a mathematician who produced ingenious and inventive ideas but whose writing required excessively hard work on the part of the reader. One can well imagine that Lagrange, now over seventy years old, receiving a poorly written memoir on a subject he had barely touched for over thirty years, was disinclined to pursue it.

In the context of the present discussion the validity of Ruffini's proof is in the end not the most important question. What matters more is that Ruffini, to a greater extent than any other writer at the end of the eighteenth-century, explored Lagrange's 'Réflexions' and their ramifications in considerable depth. Perhaps the most important consequence of Ruffini's work was that it became known in turn to Cauchy, the next major interpreter of Lagrange's 'Réflexions'.

One of Cauchy's earliest mathematical papers, presented to the Institut de France in November 1812 and published three years later, was his 'Mémoire sur le nombre des valeurs qu'une fonction peut acquérir, lorsqu'on y permute de toutes les manières possibles les quantités qu'elle renferme' ('Memoir on the number of values that a function can take when one permutes the quantities it contains in every possible way'). As the title suggests, Cauchy had taken up Lagrange's work on the permutation of variables with a view to discovering how many values a function of such variables might take. Possibly Lagrange, with Ruffini's papers and his own neglect of them still relatively fresh in his mind, had himself suggested this subject to Cauchy as a suitable research topic when Cauchy returned to Paris from Cherbourg in September 1812? There is no direct evidence for this speculation, but throughout his life Cauchy had a habit of basing some of his best work on ideas or suggestions picked up from other mathematicians. Cauchy began his paper by acknowledging both Lagrange and Vandermonde and their papers of 1771 published in Berlin and Paris,[14] but he also observed that the subject had been pursued by several Italian mathematicians, referring in particular to Ruffini's 'Risposta' to Gianfresco Malfatti, of 1805. Apart from Ruffini's compatriot Pietro Paoli, Cauchy seems to have been the only mathematician of the early nineteenth century who believed that Ruffini had succeeded in his aims. Certainly he was able to put some of Ruffini's results to good use, as will be discussed further below.

Niels Henrik Abel first started working on quintic equations around 1820 when he was 18 and still at school. As any bright young mathematician might, he tried to find a general solution, and thought for a while that he had succeeded. By 1824, however,

[13]For Ruffini's dealings with the Paris Academy see Cassinet 1988, 56–60.

[14]Lagrange (1771) [1773]; Vandermonde (1771) [1774].

he had turned to proving the impossibility of such solutions and published his first proof that year in a privately printed pamphlet.[15] He had read Cauchy's paper of 1815 and so would have known Ruffini's name but seems to have known no details of his proof. In his opening paragraph he observed that some mathematicians, by whom he presumably meant the unnamed Italians mentioned by Cauchy, had attempted to prove the impossibility of a general solution but as far as he knew no-one had yet succeeded.[16] This sounds rather more like hearsay than a careful study of the arguments. A much more direct influence on Abel than the Italian group was Lagrange, whose work he had begun to read even while he was still at school. Abel's proof, like Ruffini's, was based on the idea of roots as sums of radicals and on the number of values of a function under permutation of its variables, but unlike Ruffini he proved the crucial theorem that the coefficients of the resolvent will always be rational. This was not a problem that had troubled Lagrange or any other eighteenth-century mathematician because in their more limited experience the coefficients always *were* rational: there was therefore never any reason to suppose anything else. It was only in the proofs of Ruffini and Abel, as we have seen, that this became a critical question.

By 1828, Abel was able to be more precise about earlier work than he had been in 1824. Now he acknowledged Ruffini as the only other person to have attempted an insolvability proof, but he said he had found Ruffini's proof complicated and was not convinced of its correctness.[17] Abel's proof was not easy to follow either, however, and uncertainties continued to persist. By 1837, William Rowan Hamilton, having satisfactorily clarified some 'obscurities' in Abel's proof, declared the result correct.[18] Those who understood the matter, though, had already ceased to doubt it. Before his death in 1832 Galois had written:[19]

> Today it is a commonly held truth that general equations of degree higher than 4 cannot be solved by radicals [...] This truth has become commonly

[15]Abel 1881, I, 28–33; for a much fuller version of the proof see Abel 1881, I, 66–87.

[16]*Les géomètres se sont beaucoup occupés de la résolution générale des équations algébriques, et plusieurs d'entre eux ont cherché à en prouver l'impossibilité; mais si je ne me trompe pas, on n'y a pas réussi jusqu'à présent.* [Geometers are much concerned with the general solution of algebraic equations and several of them have sought to prove the impossibility of it; but if I am not mistaken no-one has succeeded up to the present.] Abel 1881, I, 28.

[17]*Le premier, et, si je ne me trompe, le seul qui avant moi ait cherché à démontrer l'impossibilité de la résolution algébraique des équations générales, est le géomètre Ruffini; mais son mémoire est tellement compliqué qu'il est très difficile a juger de la justesse de son raisonnement. Il me paraît que son raisonnement n'est pas toujours satisfaisant.* [The first, and, if I am not mistaken, the only person before me who has sought to prove the impossibility of algebraic solution of general equations, is the geometer Ruffini; but his memoir is so complicated that it is very difficult to judge the soundness of his reasoning. It seems to me that his reasoning is not always satisfactory.] Abel 1881, II, 218.

[18]Hamilton 1839; 1841.

[19]*C'est aujourd'hui une vérité vulgaire que les équations générales de degré supérieur au 4e ne peuvent se résoudre par radicaux, c'est-à-dire que leurs racines ne peuvent s'exprimer par des fonctions des coefficients qui ne contiendraient d'autres irrationelles que des radicaux. Cette vérité est [devenue] vulgaire [en quelque sorte par ouï dire et] quoique la plupart des géomètres en ignorent les démonstrations présentées par Ruffini, Abel, etc. démonstrations fondées sur ce qu'une telle solution est déjà impossible au cinquième degré.* Galois 1962, 33.

> known (in a way by hearsay) although most geometers are not aware of the proofs presented by Ruffini, Abel, etc.

Thus, sixty years after Lagrange began his systematic search for a method of solving higher degree equations, and three hundred years after Cardano took the first steps, the quest had finally come to an end.

Cauchy and Galois and the beginnings of group theory

In his 'Mémoire sur le nombre des valeurs qu'une fonction peut acquérir, lorsqu'on y permute de toutes les manières possibles les quantités qu'elle renferme' (1815) Cauchy began by examining functions that can take only three values. Where there are only three variables, he claimed, there are infinitely many such functions: $a_1a_2 + a_3$, $a_1(a_2 + a_3)$, and so on. One can similarly find 3-valued functions of four variables, such as $a_1a_2 + a_3a_4$ or $(a_1 + a_2)(a_3 + a_4)$. But for five variables there are neither 3-valued nor 4-valued functions, as Ruffini had shown in the two works known and cited by Cauchy, the *Teoria general delle equazioni* of 1799 and the 'Risposta' of 1805. Cauchy now set out to prove a more general theorem:[20] that the number of values of a function of n variables may be 1 or 2 but otherwise cannot be less than the greatest prime p 'contained in' (*contenu dans*) n, by which he meant the greatest prime less than or equal to n. This he proved in three stages: (i) if a function has fewer than p values then its values are not changed by any permutation of order p; (ii) if the values are not changed by any permutation of order p they are not changed by any 3-cycle; (iii) if the values are not changed by any 3-cycle, the function can take only 1 or 2 values. Cauchy ended with a stronger conjecture: that for $n \geq 5$ the number of values may be 1 or 2 but otherwise not less than n itself.

In a second memoir, immediately following the first and entitled 'Mémoire sur les fonctions qui ne peuvent obtenir que deux valeurs inégales' ('Memoir on functions which can take only two unequal values'), Cauchy explored functions of the form $(a_1 - a_2)(a_1 - a_3)\dots(a_{n-1} - a_n)$, which take just two values differing only in sign, and in doing so he essentially established the theory of determinants as alternating symmetric functions. In this context Cauchy noted the identity

$$a_2a_3^2 + a_3a_1^2 + a_1a_2^2 - a_3a_2^2 - a_2a_1^2 - a_1a_3^2 = (a_2 - a_1)(a_3 - a_1)(a_3 - a_2)$$

given by Vandermonde (see page 190), so that Vandermonde's name became associated with the determinant we would now write as

$$\begin{vmatrix} 1 & 1 & 1 \\ a_1 & a_2 & a_3 \\ a_1^2 & a_2^2 & a_3^2 \end{vmatrix}.$$

Then for thirty years Cauchy did nothing further on the number of values of a function, until in 1845 he received a paper for review from Joseph Bertrand, who had

[20] *Le nombre des valeurs différentes d'une fonction non symétrique de n quantités, ne peut s'abaisser au-dessous du plus grand nombre premier p contenu dans n, sans devenir égal à 2.* Cauchy 1815, 9.

JOURNAL
DE L'ÉCOLE POLYTECHNIQUE.

MÉMOIRE

Sur le Nombre des Valeurs qu'une Fonction peut acquérir ; lorsqu'on y permute de toutes les manières possibles les quantités qu'elle renferme ;

PAR A. L. CAUCHY, INGÉNIEUR DES PONTS ET CHAUSSÉES.

MM. *Lagrange* et *Vandermonde* sont, je crois, les premiers qui aient considéré les fonctions, de plusieurs variables relativement au nombre de valeurs qu'elles peuvent obtenir, lorsqu'on substitue ces variables à la place les unes des autres. Ils ont donné plusieurs théorèmes intéressans relatifs à ce sujet, dans deux mémoires imprimés en 1771, l'un à Berlin, l'autre à Paris. Depuis ce temps, quelques géomètres italiens se sont occupés avec succès de cette matière, et particulièrement M. *Ruffini*, qui a consigné le résultat de ses recherches dans le tome XII des Mémoires de la Société italienne, et dans sa Théorie des équations numériques. Une des conséquences les plus remarquables des travaux de ces divers géomètres, est qu'avec un nombre donné de lettres on ne peut pas toujours former une fonction qui ait un nombre déterminé de valeurs. Les caractères par lesquels cette

XVII.e Cahier. A

An early paper from Cauchy (1815), referring to Lagrange, Vandermonde, and Ruffini.

done further work on Cauchy's conjecture of 1815. Cauchy's interest was immediately rekindled and between September 1845 and April 1846 he published a stream of papers on permutations of the variables of a function.[21] These new investigations led him almost immediately to the concept of a 'system of combined substitutions' (*système des substitutions conjuguées*), or what is now known as a 'group'.

The term 'group' (in French, *groupe*) had meanwhile been coined independently by Galois some fifteen years earlier. Galois too had examined systems of permutations but unlike Cauchy was not interested in them simply from a combinatorial point of view. Galois' aim was to determine which equations *were* algebraically solvable, and so for him, as for Lagrange, the variables he was concerned with were roots of equations. What Galois found was that the solvability of a given equation depends upon the structural properties of the group of permutations of its roots, now known as its Galois group. His work was tragically cut short by his untimely and needless death in 1832 at the age of 20, but his papers were eventually published by Joseph Liouville in 1846. By that time Cauchy had also published his long run of papers in the *Comptes rendus*.

By the early 1850s it became clear that Cauchy and Galois, though following different approaches, had discovered similar algebraic structures: what Galois called a *groupe* and Cauchy called a *système des substitutions*. The theory was further and rapidly consolidated after 1857 when the Paris Academy set the problem of discovering the number of values of a function under permutation of its variables as the subject of its Grand Prix for 1860. The competition elicited entries from Thomas Kirkman, Émile Mathieu, and Camille Jordan, though the prize itself was never awarded. Cauchy was on the committee that set the subject, and in the statement of the problem one recognizes precisely the area of research Lagrange had opened up 90 years earlier and which he himself may have recommended to Cauchy right at the beginning of Cauchy's mathematical career.

Thus all the writers described in this section, Ruffini, Abel, Cauchy, and Galois, took Lagrange's ideas of 1771–72 as their starting point. Their motivations differed: Ruffini and Abel wanted to prove that quintic equations were not algebraically solvable; Galois hoped to determine which equations of degree higher than four *could* be solved; Cauchy was interested in the raw question of the number of values of functions under permutations of their variables. All four of them, as indeed had Lagrange himself, arrived at concepts and theorems that later became absorbed into the theory of groups. Thus the old algebra of equation solving was transformed in the early nineteenth century into a quite different kind of algebra, now usually described as 'abstract' or 'modern'.

When Cardano in 1545 turned his attention to the problem of transforming equations without actually solving them, he too had been engaging in a process of generalization, from particular techniques of solution to a more all-embracing vision of equations as mathematical objects in their own right. In the centuries between Cardano and Lagrange, algebra took on a variety of names, forms, and applications, but always one of its characteristic features was the process of increasing abstraction from one level of thinking to another. Cardano, in embarking on that path, transformed not just

[21] For the details see Neumann 1989.

equations but algebra itself. Lagrange two centuries later looked back to Cardano and his successors, and in doing so he too produced ideas that were again to change the nature and scope of algebra, this time from the study of equations to the investigation of the abstract structures that later became known as groups. Lagrange rightly recognized Cardano's work as the beginning of a key period in algebra; his own work in turn initiated another.

Bibliography

Dates. Several of the journal articles in the bibliography are listed with multiple dates. A date immediately following a title is the year when the paper is known to have been written (as recorded, for instance, in the register of an Academy). Dates in round brackets are years for which a volume of papers was published. Dates in square brackets are years in which the volume actually appeared. Thus, Euler's 'Recherches sur les racines imaginaires des equations' was presented to the Berlin Academy in 1746. It was later included in the volume of *Mémoires* for the year 1749, which was eventually printed in 1751.

Translations. English translations where they are known to exist are noted alongside the original text (Witmer's 1968 translation of Cardano's *Ars magna*, for instance). References to the translations are given in the footnotes after references to the original in cases where the reader may not easily be able to consult the primary text, but without comment on the accuracy or otherwise of such translations.

Euler. Euler's books, papers, and some letters were catalogued in roughly chronological order by Gustav Eneström in the early twentieth century. Eneström numbers are given in square brackets thus: [E170] after each Euler reference. All Euler's published papers, and some translations, are accessible in the Euler Archive at http://www.math.dartmouth.edu/~euler/

Web resources. The number of sources available online is increasing so rapidly that it is impossible to give a comprehensive list: it is always worth searching for new additions. Sites that have been particularly useful in relation to the material in this book are:

European Cultural Heritage Online (ECHO): http://echo.mpiwg-berlin.mpg.de/home

English books up to 1700 (EEBO): http://eebo.chadwyck.com/home

English books 1700–1800 (ECCO): http://find.galegroup.com/ecco/

French books (Gallica): http://gallica.bnf.fr

Publications of the Berlin Academy:
http://bibliothek.bbaw.de/bibliothek/digital/index.html

Publications of the Paris Academy:
http://www.academie-sciences.fr/archives/ressources_bnf.htm

Publications by Cardano: http://www.cardano.unimi.it/testi/opera.html
Publications by Euler: http://www.math.dartmouth.edu/~euler/

Abel, Niels Henrik, *Oeuvres complètes de Niels Henrik Abel* (second edition), edited by Ludvig Sylow and Sophus Lie, 2 vols, Christiania, 1881.

d'Alembert, Jean le Rond, 'Rechérches sur le calcul intégral', *Mémoires de l'Académie Royale des Sciences et Belles Lettres de Berlin*, 2 (1746) [1748], 182–224.

Anonymous [probably Leibniz], 'Treatise of algebra both historical and practical, with some additional treatises, by Iohan Wallis', *Acta eruditorum*, 5 (1686), 283–289 [last page misnumbered 489].

Anonymous, *The Georgian era: memoirs of the most eminent persons, who have flourished in Great Britain, from the access of George the First to the demise of George the Fourth*, 4 vols, London, 1832–34.

Aubrey, John, *Brief lives, chiefly of contemporaries, set down by John Aubrey between the years 1669 and 1696*, 2 vols, edited by Andrew Clark, Oxford, 1898.

Ayoub, Raymond G, 'Paolo Ruffini's contributions to the quintic', *Archive for history of exact sciences*, 23 (1980), 253–277.

Barrow, Isaac, *Lectiones geometricae: in quibus (praesertim) generalia curvarum linearum symptomata declarantur*, London, 1670.

Barrow-Green, June, 'From cascades to calculus: Rolle's theorem' in *The Oxford handbook of the history of mathematics*, Eleanor Robson and Jacqueline Stedall (eds), Oxford University Press, 2009.

Bashmakova, Isabella, and Galina Smirnova, *The beginnings and evolution of algebra*, The Mathematical Association of America, 2000.

de Beaune, Florimond, 'De aequationum natura, constitutione, et limitibus opuscula duo', in Descartes 1659–61, II, 49–152.

Beeley, Philip, and Christoph J Scriba (eds), *Correspondence of John Wallis (1616–1703)*, 2 vols to date, Oxford University Press, 2003–.

Beery, Janet, and Jacqueline Stedall, *Thomas Harriot's doctrine of triangular numbers: the 'Magisteria magna'*, European Mathematical Society, 2009.

Bellhouse, David, 'Decoding Cardano's *Liber de ludo aleae*', *Historia mathematica*, 32 (2005), 180–202.

Bertrand, Joseph, 'Mémoire sur le nombre de valeurs que peut prendre une fonction quand on y permute les lettres qu'elle renferme (Extrait)', *Comptes rendus de l'Académie des Sciences*, 20 (1845), 798–700.

Bertrand, Joseph, 'Mémoire sur le nombre de valeurs que peut prendre une fonction quand on y permute les lettres qu'elle renferme', *Journal de l'École Polytechnique* (Cahier 30), 18 (1848), 123–140.

Bezout, Étienne, 'Mémoire sur plusieurs classes d'équations de tous les degrés qui admettent une solution algébrique', *Mémoires de l'Académie Royale des Sciences de Paris*, (1762) [1764], 17–52.

Bezout, Étienne, 'Rechérches sur le degré des équations résultantes de l'évanouissement des inconnues', *Mémoires de l'Académie Royale des Sciences de Paris*, (1764) [1767], 288–338.

Bezout, Étienne, 'Mémoire sur la résolution générale des équations de tous les degrés', (1763), *Mémoires de l'Académie Royale des Sciences de Paris*, (1765) [1768], 533–552.

Bezout, Étienne, (A) *Cours de mathématiques, à l'usage des Gardes du Pavillon et de la Marine*, 6 vols, Paris, 1764–66, 1770–72, 1781–84, 1787–89; second edition: *Cours de mathématiques, à l'usage des Gardes du Pavillon et de la Marine et des éleves de l'École Polytechnique*, 6 vols, Paris, 1798–99, 1800–03; third edition: *Cours de mathématiques, à l'usage de la Marine, et des élèves de l'École Polytechnique*, augmented by Garnier, 1809. A parallel publication was (B) *Cours de mathématiques, à l'usage du Corps royal de l'Artillerie*, 4 vols, Paris, 1770–72, 1781, 1788–90; second edition 1797. Versions (A) and (B) were combined as *Cours de mathématiques, à l'usage de la Marine et de l'Artillerie*, with notes by Reynaud, 1822, 1828—29.

Bezout, Étienne, *Théorie générale des équations algébriques*, Paris, 1779; translated by Eric Feron as *General theory of algebraic equations*, Princeton University Press, 2006.

Bombelli, Rafael, *L'algebra parte maggiore dell'aritmetica divisa in tre libri*, Bologna, 1572; reprinted as *L'algebra*, Bologna, 1579.

Bos, Henk J M, *Redefining geometrical exactness: Descartes' transformation of the Early Modern concept of construction*, Springer-Verlag, 2001.

Bring, Erland Samuel, *Meletemata quaedam mathematica circa transformationem aequationum algebraicarum*, Lund, 1786.

Bryce, Robert A, 'Paolo Ruffini and the quintic equation', *Symposia mathematica*, 27 (1986), 169–185.

Campbell, George, 'A method for determining the number of impossible roots in adfected aequations', *Philosophical Transactions of the Royal Society*, 35 (July 1727–1728), 515–531.

Campbell, George, *Remarks on a paper published by Mr. MacLaurin, in the Philosophical Transactions for the Month of May, 1729*, Edinburgh, 1729.

Cardano, Girolamo, *Artis magnae, sive, de regulis algebraicis, liber unus* (= *Ars magna*), Nuremberg, 1545; translated by T Richard Witmer as *Ars magna, or, the rules of algebra*, MIT Press, 1968; reprinted Dover Publications, 1993.

Cardano, Girolamo, *Opus novum de proportionibus numerorum, motuum, ponderum, sonorum, aliarumque rerum mensurandarum,* [...] *Praeterea, Artis magnae, sive de regulis algebraicis, liber unus, abstrusissimus et inexhaustus plane totius arithmeticae thesaurus ab authore recens multis in locis recognitus et auctus. Item, De aliza regula liber*, Basel, 1570.

Cardano, Girolamo, *Opera omnia*, 10 vols, Leiden, 1663.

Cardano, Girolamo, *The book of my life* (De vita propria liber), J M Dent, 1931; reprinted New York Review Books, 2002.

Cassinet, Jean, 'Paolo Ruffini (1765–1822): la résolution algébrique des équations et les groupes de permutations', *Bollettino di Storia delle Scienze Matematiche*, 8 (1988), 21–69.

Castillon, Johann, *Arithmetica universalis, sive de compositione et resolutione arithmetica. Cum commentario Johannis Castillionei*, Amsterdam, 1761.

Castillon, Johann, 'Mémoire sur les équations résolues par M. de Moivre, avec quelques réflexions sur ces équations et sur les cas irréducibles', *Nouveaux Mémoires de l'Académie Royale des Sciences et Belles Lettres de Berlin*, 2 (1771) [1773], 254–272.

Cauchy, Augustin-Louis, 'Sur le nombre des valeurs qu'une fonction peut acquérir, lorsqu'on y permute de toutes les manières possibles les quantités qu'elle renferme' (1812), *Journal de l'École Polytechnique* (Cahier 17), 10 (1815a), 1–28; and in Cauchy, *Oeuvres* (2), I, 64–90.

Cauchy, Augustin-Louis, 'Sur les fonctions qui ne peuvent obtenir que deux valeurs inégales et de signes contraires par suite des transpositions opérées entre les variables qu'elles renferment' (1812), *Journal de l'École Polytechnique* (Cahier 17), 10 (1815b), 29–112; and in Cauchy, *Oeuvres* (2), I, 91–169.

Céu Silva, Maria, 'The algebraic content of Bento Fernandes's *Tratado da arte de arismetica* (1555)', *Historia mathematica*, 35 (2008), 190–219.

Collins, John, 'An account concerning the resolution of equations in numbers', *Philosophical Transactions of the Royal Society*, 4 (1669), 929–934.

Collins, John, 'A letter from Mr John Collins [...] giving his thoughts about some defects in algebra', *Philosophical Transactions of the Royal Society*, 14 (1684), 575–582.

Collins, John, 'Narrative about aequations' (Part I), 1670, in Gregory 1939, 113–117.

Colson, John, 'Aequationum cubicarum et biquadraticarum, tum analytica, tum geometrica et mechanica, resolutio universalis', *Philosophical Transactions of the Royal Society*, 25 (1707), 2353–2368.

Cramer, Gabriel, *Introduction à l'analyse des lignes courbes algébriques*, Geneva, 1750.

Derbyshire, John, *Unknown quantity: a real and imagined history of algebra*, The Joseph Henry Press, 2006; Atlantic Books, 2007.

Descartes, René, *La géométrie*, appended to *Discours de la méthode*, Leiden, 1637; reprinted and translated in *The geometry of René Descartes*, edited by David Eugene Smith and Marcia L Latham, Open Court, 1925; reprinted Dover Publications, 1954.

Descartes, René, *Geometria*, translated by Frans van Schooten, Leiden, 1649.

Descartes, René, *Geometria*, edited by Frans van Schooten, 2 vols, Amsterdam, 1659–61.

Digges, Thomas, *An arithmeticall militare treatise, named Stratioticos: compendiously teaching the science of nu[m]bers, as well in fractions as integers, and so much of the rules and aequations algebraicall and arte of numbers cossicall, as are requisite for the profession of a soldiour*, London, 1579.

Dionis du Séjour, Achille-Pierre, and Mathieu-Bernard Goudin, *Traité des courbes algébriques*, Paris, 1756.

Dulaurens, Francis, *Specimina mathematica*, Paris, 1667.

Dulaurens, Francis, 'Responsio ... ad epistolam d. Wallisii ad clarissimum virum Oldenburgium scriptam', printed pamphlet, 1668. A copy of this pamphlet is bound with Wallis's copy of Dulaurens' *Specimina mathematica* in the Bodleian Library, Oxford (Savile G.8).

van Egmond, Warren, 'The earliest vernacular treatment of algebra: the *Libro di ragioni* of Paolo Gerardi [1328]', *Physis*, 20 (1978), 155–189.

van Egmond, Warren, 'The algebra of Master Dardi of Pisa', *Historia mathematica*, 10 (1983), 399–421.

Euler, Leonhard, 'De formis radicum aequationum cuiusque ordinis coniectatio', (1733), *Commentarii Academiae Scientiarum Petropolitanae*, 6 (1732–33) [1738], 216–231. [E30]

Euler, Leonhard, 'Sur un contradiction apparente dans la doctrine des lignes courbes', (1747), *Mémoires de l'Académie Royale des Sciences et Belles Lettres de Berlin*, 4 (1748a) [1750], 219–233. [E147]

Euler, Leonhard, 'Demonstration sur le nombre des points, ou deux lignes des ordres quelconques peuvent se couper', (1748), *Mémoires de l'Académie Royale des Sciences et Belles Lettres de Berlin*, 4 (1748b) [1750], 234–248. [E148]

Euler, Leonhard, *Introductio ad analysin infinitorum*, 2 vols, Lausanne, 1748.

Euler, Leonhard, 'Recherches sur les racines imaginaires des equations', (1746), *Mémoires de l'Académie Royale des Sciences et Belles Lettres de Berlin*, 5 (1749) [1751], 222–288. [E170]

Euler, Leonhard, 'Demonstratio gemina theorematis Neutoniani, quo traditur relatio inter coefficientes cuiusvis aequationis algebraicae et summas potestatum radicum eiusdem', (1747), *Opuscula varii argumenti*, 2 (1750), 108–120. [E153]

Euler, Leonhard, *Institutiones calculi differentialis*, (1748), St Petersburg, 1755. [E212]

Euler, Leonhard, 'De resolutione aequationum cuiusvis gradus', (1753), *Novi commentarii Academiae Scientiarum Petropolitanae*, 9 (1762–63) [1764], 70–98. [E282]

Euler, Leonhard, 'Nouvelle méthode d'éliminer les quantités inconnues des equations', (1752), *Mémoires de l'Académie Royale des Sciences et Belles Lettres de Berlin*, 20 (1764) [1766], 91–104. [E310]

Euler, Leonhard, 'Nova criteria radices aequationum imaginarias dignoscendi', (1767), *Novi commentarii Academiae Scientiarum Petropolitanae*, 13 (1768) [1769], 89–119. [E370]

Euler, Leonhard, *Universal'naya arifmetika*, 2 vols, St Petersburg, 1768–69, 1787–88; translated into German as *Vollständige Anleitung zur Algebra*, 2 vols, St Petersburg, 1770, Lund, 1771; translated into French as *Elémens d'algèbre* by Jean Bernoulli with additions by Lagrange, 2 vols, Lyon, 1774, 1795; retranslated from French as *Vollständige Anleitung zur niedern und höhern Algebra* by Johann Philipp Grüson, Berlin, 1796–97; retranslated from French as *Nachal'naya osnovaniya algebry*, St Petersburg, 1798; translated into English as *Elements of algebra* by Francis Horner, London, 1797, reprinted Springer-Verlag, 1972.

Euler, Leonhard, 'Observationes circa radices aequationum', (1770), *Novi commentarii Academiae Scientiarum Petropolitanae*, 15 (1770) [1771], 51–74. [E406]

Euler, Leonhard, 'De serie Lambertina plurimisque eius insignibus proprietatibus', (1776), *Acta Academiae Scientarum Imperialis Petropolitanae*, (1779) [1783], 29–51. [E532]

Euler, Leonhard, 'Analysis facilis et plana ad eas series maxime abstrusas perducens, quibus omnium aequationum algebraicarum non solum radices ipsae sed etiam quaevis earum potestates exprimi possunt', *Nova acta Academiae Scientarum Imperialis Petropolitanae*, 4 (1789a), 55–73. [E631]

Euler, Leonhard, 'De innumeris generibus serierum maxime memorabilium, quibus omnium aequationum algebraicarum non solum radices ipsae sed etiam quaecumque earum potestates exprimi possunt', *Nova acta Academiae Scientarum Imperialis Petropolitanae*, 4 (1789b), 74–95. [E632]

Euler, Leonhard, 'Innumerae aequationum formae ex omnibus ordinibus, quarum resolutio exhiberi potest', *Nova acta Academiae Scientarum Imperialis Petropolitanae*, 6 (1790), 25–35. [E644]

Euler, Leonhard, 'Methodus nova ac facilis omnium aequationum algebraicarum radices non solum ipsas sed etiam quascumque earum potestates per series concinnas exprimendi', *Nova acta Academiae Scientarum Imperialis Petropolitanae* 12 (1801), 71–90. [E711]

Franci, Rafaella, and Laura Toti Rigatelli, 'Towards a history of algebra from Leonardo of Pisa to Luca Pacioli', *Janus*, 72 (1985), 17–82.

Galois, Évariste, *Écrits et mémoires mathématiques*, edited by Robert Bourgne and Jean-Paul Azra, Gauthier-Villars, 1962.

Girard, Albert, *Invention nouvelle en l'algebre*, Leiden, 1629.

'sGravesande, Willem, *Arithmetica universalis, sive de compositione et resolutione arithmetica liber*, Leiden, 1732.

Gregory, James, *James Gregory: Tercentenary memorial volume*, edited by H W Turnbull, Royal Society of Edinburgh, 1939.

de Gua de Malves, Jean Paul, *Usages de l'analyse de Descartes*, Paris, 1740.

de Gua de Malves, Jean Paul, 'Démonstrations de la régle de Descartes, pour connoître le nombre des Racines positives et négatives dans les équations qui n'ont point de Racines imaginaires', *Mémoires de l'Académie Royale des Sciences de Paris*, (1741a) [1744], 72–96.

de Gua de Malves, Jean Paul, 'Recherche du nombre des racines réelles ou imaginaires, réelles positives ou réelles négatives, qui peuvent se trouver dans les équations de tous les degrés', *Mémoires de l'Académie Royale des Sciences de Paris*, (1741b) [1744], 435–494.

Hall, Rupert A, and Maria Boas Hall (eds), *The correspondence of Henry Oldenburg*, 13 vols, University of Wisconsin Press, 1965–86.

Halley, Edmund, 'Methodus nova accurata et facilis inveniendi radices aequationum quarumcumque generaliter, sine praevia reductione', *Philosophical Transactions of the Royal Society*, 18 (1694), 136–148.

Hamilton, Wiliam Rowan, 'On the argument of Abel, respecting the Impossibility of expressing a root of any general equation above the fourth degree, by any finite combination of radicals and rational functions' (1837), *Transactions of the Royal Irish Academy*, 18 (1839), 171–259; for an abstract of this paper see 'Investigations respecting equations of the fifth degree', *Proceedings of the Royal Irish Academy*, 1 (1841), 76–80.

Harriot, Thomas, unpublished manuscripts: British Library, Add MSS 6782–6789, and Sussex Public Record Office, Petworth HMC MSS 240–241.

Harriot, Thomas, *Artis analyticae praxis*, London, 1631.

Harriot, Thomas, *The greate invention of algebra: Thomas Harriot's treatise on equations*, edited and translated by Jacqueline Stedall, Oxford University Press, 2003.

Harriot, Thomas, *Thomas Harriot's Artis analyticae praxis: an English translation with commentary*, edited and translated by Muriel Seltman and Robert Goulding, Springer-Verlag, 2007.

Hérigone, Pierre, *Cursus mathematicus*, 6 vols, Paris, first edition 1634–42; second edition 1644.

Høyrup, Jens, *Jacopo da Firenze's* Tractatus algorismi *and early Italian abbacus culture*, Springer-Verlag, 2007.

Hudde, Jan, 'De reductione aequationum', in Descartes 1659–61, I, 407–506.

Hutton, Charles, *A mathematical and philosophical dictionary*, 2 vols, London, 1795–96.

Hutton, Charles, *Tracts on mathematical and philosophical subjects*, 3 vols, London, 1812.

Kästner, Abraham Gotthelph, 'Demonstratio theorematis Harriotti', appended to Johann Castillon, *Arithmetica universalis*, Amsterdam, 1761, 118–123.

Katz, Victor J, *A history of mathematics: an introduction*, third edition, Addison-Wesley, 2009.

Kinckhuysen, Gerard, *Algebra, ofte stel-konst*, Haarlem, 1661.

Kline, Morris, *Mathematical thought from ancient to modern times*, Oxford University Press, 1972; reprinted in 3 vols, 1990.

Lagrange, Joseph-Louis, 'Sur la résolution des équations numériques', (1769), *Mémoires de l'Académie royale des Sciences et Belles Lettres de Berlin*, 23 (1767) [1769], 311–352.

Lagrange, Joseph-Louis, 'Additions au mémoire sur la résolution des équations numériques', (1769, 1770), *Mémoires de l'Académie royale des Sciences et Belles Lettres de Berlin*, 24 (1768) [1770], 111–180.

Lagrange, Joseph-Louis, 'L'élimination des inconnues dans les équations, (1767), *Mémoires de l'Académie Royale des Sciences et Belles Lettres de Berlin*, 25 (1769) [1771], 303–318.

Lagrange, Joseph-Louis, 'Réflexions sur la résolution algébrique des équations' (1771, 1772), *Nouveaux Mémoires de l'Académie Royale des Sciences et Belles Lettres de Berlin*, 1 (1770) [1772], 134–172; 173–215, and 2 (1771) [1773], 138–189; 189–253; and in *Oeuvres*, III, 203–421.

Lagrange, Joseph-Louis, 'Recherches sur la détermination des racines imaginaires dans les équations litérales', (1772 and 1777), *Nouveaux Mémoires de l'Académie Royale des Sciences et Belles Lettres de Berlin*, 8 (1777) [1779], 111–139.

Lagrange, Joseph-Louis, *Théorie des fonctions analytiques*, Paris, 1797.

Lagrange, Joseph-Louis, *Traité de la résolution des équations numériques de tous les degrés*, Paris, 1808.

Lalande, Jérôme, 'Notice sur la vie de Condorcet', *Mercure de France*, 20 January 1796, 143.

Leibniz, Gottfried Wilhelm, *Der Briefwechsel von Gottfried Wilhelm Leibniz mit Mathematikern*, edited by C I Gerhardt, Berlin, 1899.

Leibniz, Gottfried Wilhelm, *Sämtliche Schriften und Briefe*, series 3: 'Mathematischer naturwissenschaftlicher und technischer Briefwechsel', Deutsche Akademie der Wissenschaften zu Berlin, 1976–.

Leonardo Pisano, *Fibonacci's Liber abaci: Leonardo Pisano's book of calculation*, translated by L E Sigler, Springer-Verlag, 2002.

Livio, Mario, *The equation that couldn't be solved: how mathematical genius discovered the language of symmetry*, Simon and Schuster, 2005; Souvenir Press, 2006.

Maclaurin, Colin, 'A letter from Mr Colin Maclaurin [...] concerning aequations with impossible roots', (1726), *Philosophical Transactions of the Royal Society*, 34 (1726–June 1727), 104–112.

Maclaurin, Colin, 'A second letter from Mr Colin Maclaurin [...] concerning the roots of equations, with the demonstration of other rules in algebra', (1729), *Philosophical Transactions of the Royal Society*, 36 (1729–30), 59–96.

Maclaurin, Colin, *A defence of the letter published in the Philosophical Transactions for March and April 1729 concerning the impossible roots of equations; in a letter from the author to a friend at London', 1730*, Edinburgh, 1730; reprinted in Mills 1982, 222–241.

Maclaurin, Colin, *A treatise of algebra*, London, 1748.

Mahoney, Michael Sean, *The mathematical career of Pierre de Fermat 1601–1665*, Princeton University Press, 1973; second edition, 1994.

Malcolm, Noel and Jacqueline Stedall, *John Pell (1611–1685) and his correspondence with Sir Charles Cavendish: the mental world of an early modern mathematician*, Oxford University Press, 2005.

Maseres, Francis, 'A method of extending Cardan' rule for resolving one case of a cubick equation of this form, $x^3 * -qx = r$, to the other case of the same equation, which it is not naturally fitted to solve, and which is therefore often called the irreducible case', *Philosophical Transactions of the Royal Society*, 68 (1778), 902–949.

Mills, Stella (ed), *The collected letters of Colin MacLaurin*, Shiva Publishing, 1982.

de Moivre, Abraham, 'Aequationum quarundam potestatis tertiae, quintae, septimae, nonae, et superiorum, ad infinitum usque pergendo, in terminis finitis, ad instar regularum pro cubicis quae vocantur Cardani, resolutio analytica', *Philosophical Transactions of the Royal Society*, 25 (1707), 2368–2371.

de Moivre, Abraham, *Miscellanea analytica de seriebus et quadraturis*, London, 1730.

Montucla, Étienne, *Histoire des mathématiques*, 2 vols, Paris, 1758.

Neumann, Peter M, 'On the date of Cauchy's contributions to the founding of the theory of groups', *Bulletin of the Australian Mathematical Society*, 40 (1989), 293–302.

Newton, Isaac, *Opticks: or, a treatise of the reflexions, refractions, inflexions and colours of light. Also two treatises of the species and magnitude of curvilinear figures*, London, 1704.

Newton, Isaac, *Arithmetica universalis; sive de compositione et resolutione arithmetica liber*, Cambridge, 1707; second Latin edition, London, 1722.

Newton, Isaac, *Universal arithmetick: or, a treatise of arithmetical composition and resolution*, London, 1720; second English edition, London 1728.

Newton, Isaac, *The correspondence of Isaac Newton*, edited by H W Turnbull, 7 vols, Cambridge, 1959–77.

Newton, Isaac, *The mathematical papers of Isaac Newton*, edited by D T Whiteside, 8 vols, Cambridge, 1967–81.

Nicole, François, 'Sur le cas irreducible du troisième degré', *Mémoires de l'Académie Royale des Sciences de Paris*, (1738) [1740], 97–102.

Nový, Luboš, *Origins of modern algebra*, Noordhoff International Publishing, 1973

Oughtred, William, *The key of mathematics new filed*, London, 1647.

Ore, Øystein, *Cardano, the gambling scholar*, Princeton University Press, 1953; reprinted Dover Publications, 1965.

Pacioli, Luca, *Summa de arithmetica, geometria, proportioni et proportionalita*, Venice, 1494.

Paoli, Pietro, *Elementi d'algebra*, 2 vols, Pisa, 1794.

Paoli, Pietro, *Supplemento agli elementi d'algebra*. Pisa, 1804.

Pesic, Peter, 'François Viète, father of modern cryptoanalysis: two new manuscripts', *Cryptologia*, 21 (1997a), 1–29

Pesic, Peter, 'Secrets, symbols, and systems: parallels between cryptanalysis and algebra, 1580–1700', *Isis*, 88 (1997b), 674–692.

Prestet, Jean, *Nouveaux elemens des mathematiques*, Paris, third edition, 2 vols, 1694.

Raphson, Joseph, *Analysis aequationum universalis, seu ad equationes algebraicas resolvendas methodus generalis et expedita ex nova infinitarum serierum methodo deducta et demonstrata*, London, 1690; 1697; 1702.

Reyneau, Charles René, *Analyse demontrée ou la methode de resoudre les problêmes des mathématiques, et d'apprendre facilement ces sciences*, 2 vols, Paris, 1708.

Rigaud, Stephen Jordan (ed), *Correspondence of scientific men of the seventeenth century*, 2 vols, Oxford, 1841.

Rolle, Michel, *Traité d'algebre ou principes généraux pour resoudre les questions de mathématique*, Paris, 1690.

Rolle, Michel, *Démonstration d'une méthode pour résoudre les égalitez de tous les degrez*, Paris, 1691.

Ronan, Mark, *Symmetry and the monster: one of the greatest quests of mathematics*, Oxford University Press, 2006.

Ruffini, Paolo, *Teoria general delle equazioni in cui si demostra impossibile la soluzione algebrica delle equazioni generali di grado superiore al quarto*, 2 vols, Bologna, 1799.

Ruffini, Paolo, 'Risposta ...ai dubbi propostigli dal socio Gianfresco Malfatti sopre la insolubilità delle equazioni di grado superiore al quarto', *Memorie di Matematica e di Fisica della Società Italiana delle Scienze*, 12 (1805), 213–267.

Ruffini, Paolo, *Riflesioni intorno alla soluzione delle equazioni algebraiche*, Modena, 1813.

Saunderson, Nicholas, *Elements of algebra*, 2 vols, Cambridge, 1740.

du Sautoy, Marcus, *Finding moonshine*, Fourth Estate, 2008.

von Segner, Johann Andreas, 'Démonstration de la régle de Descartes, pour connoitre le nombre des racines affirmatives et négatives qui peuvent se trouver dans les 'equations', *Mémoires de l'Académie Royale des Sciences et Belles Lettres de Berlin*, 12 (1756) [1758], 292–299.

Scriba, Christoph J, 'Mercator's Kinckhuysen translation in the Bodleian Library at Oxford', *British Journal for the History of Science*, 2 (1964), 45–58.

Simpson, Thomas, *The doctrine and application of fluxions*, 2 vols, London, 1750.

Stedall, Jacqueline, *A discourse concerning algebra: English algebra to 1685*, Oxford University Press, 2002.

Stedall, Jacqueline, 'Symbolism, combinations, and visual imagery in the mathematics of Thomas Harriot', *Historia mathematica*, 34 (2007), 380–401.

Stevin, Simon, *L'arithmetique* [...] *aussi l'algebre, avec les equations de cinc quantitez*, Leiden, 1585; reprinted in *The principal works of Simon Stevin*, Amsterdam, 1958, vol IIB, 477–708.

Stevin, Simon, *Les oeuvres mathématiques*, edited by Albert Girard, Leiden, 1634.

Stewart, Ian, *Why beauty is truth: a history of symmetry*, Basic Books, 2007.

Stifel, Michael, *Arithmetica integra*, Nurenberg, 1544.

Stillwell, John, *Mathematics and its history*, Springer-Verlag, 2000; second edition 2002.

Stirling, James, *Lineae tertii ordinis Neutonianae, sive illustratio tractatus D. Neutoni de enumeratione linearum tertii ordinis*, Oxford, 1717; reprinted Paris, 1797.

Struik, Dirk J, *A concise history of mathematics*, G Bell and Sons, 1954.

Tartaglia, Niccolò, *Quesiti, et inuentioni diuerse*. Venice, 1546.

Thomas, David J, 'Raphson, Joseph (*fl.* 1689–1712)', *Oxford dictionary of national biography*, Oxford University Press, 2004.

von Tschirnhaus, Ehrenfried Walter, 'Nova methodus auferendi omnes terminos intermedios ex data aequatione', *Acta eruditorum*, (1683), 204–207.

Vandermonde, Alexandre-Théophile, 'Mémoire sur la resolution des équations' (1770), *Mémoires de l'Académie Royale des Sciences à Paris*, (1771a) [1774], 365–416.

Vandermonde, Alexandre-Théophile, 'Remarques sur des problèmes de situation', *Mémoires de l'Académie Royale des Sciences à Paris*, (1771b) [1774], 566–574.

Vandermonde, Alexandre-Théophile, 'Mémoire sur des irrationelles de différents ordres avec une application au cercle', *Mémoires de l'Académie Royale des Sciences à Paris*, (1772a) [1775], 489–498.

Vandermonde, Alexandre-Théophile, 'Mémoire sur élimination' (1771), *Mémoires de l'Académie Royale des Sciences à Paris*, (1772b) [1776], 516–532.

Viète, François, *In artem analyticem isagoge*, Tours, 1591.

Viète, François, *Zeteticum libri quinque*, Tours, 1591 or 1593.

Viète, François, *Effectionum geometricarum canonica recensio*, Tours, 1593.

Viète, François, *Supplementum geometriae*, Tours, 1593.

Viète, François, *Responsum ad problema quod* [...] *proposuit Adrianus Romanus*, Paris, 1595.

Viète, François, *De numerosa potestatum ad exegesin resolutione*, Paris, 1600.

Viète, François, *De recognitione et emendatione aequationum tractatus duo*, Paris, 1615.

Viète, François, *Ad angularium sectionum analyticen theoremata*, edited and completed by Alexander Anderson, Paris, 1615.

Viète, François, *Opera mathematica*, edited by Francis van Schooten, Leiden, 1646; reprinted in facsimile by Georg Olms, 2001; partially translated by T Richard Witmer in *The analytic art: nine studies in algebra, geometry and trigonometry* [...] *by François Viete*, The Kent State University Press, 1983.

van der Waerden, Bartel L, *A history of algebra from al-Khwārizmī to Emmy Noether*, Springer-Verlag, 1980.

Wallis, John, *Arithmetica infinitorum*, Oxford, 1656; translated as *The arithmetic of infinitesimals* by Jacqueline Stedall, Springer-Verlag, 2004.

Wallis, John, 'Concerning some mistakes of a book entitled Specimina mathematica Francisci Dulaurens, especially touching a certain probleme, affirm'd to have been proposed by Dr. Wallis to the mathematicians of all Europe, for a solution', *Philosophical Transactions of the Royal Society*, 3 (1668a), 654–655.

Wallis, John, 'Some animadversions, written in a letter by Dr. John Wallis, on a printed paper, entitul'd Responsio Francisci du Laurens ad epistolam D. Wallisii ad Cl. V. Oldenburgium scriptam', *Philosophical Transactions of the Royal Society*, 3 (1668b), 744–750.

Wallis, John, 'A second letter of Dr. John Wallis on the same printed paper of Francisus Du Laurens, mention'd in the next foregoing Transactions', *Philosophical Transactions of the Royal Society*, 3 (1668c), 775–779.

Wallis, John, *A treatise of algebra historical and practical*, London, 1685.

Waring, Edward, *Miscellanea analytica de aequationibus algebraicis, et curvarum proprietatibus*, Cambridge, 1762.

Waring, Edward, *Meditationes algebraicae*, Cambridge, 1770, 1782.

Waring, Edward, 'Original letter of the late Dr. Waring to the Rev. Dr. Maskelyne', *The monthly magazine*, 7 (1799), 306–310.

Waring, Edward, *Meditationes algebraicae: an English translation of the work of Edward Waring*, translated by Dennis Weeks, American Mathematical Society, 1991.

von Wolff, Christian, *Elementa matheseos universalis*, 2 vols, Halle, 1713; translated into English as *A treatise of algebra; with the application of it to a variety of problems in arithmetic, to geometry, trigonometry, and conic sections*, translator unknown, London, 1739.

Index